ADVANTAGE Math

7

MW00723567

Table of Contents

Table of Contents

CREDITS
Concept Development: Kent Publishing Services, Inc.
Written by: Tom Hatch
Editor: Dawn Purney
Designer: Moonhee Pak
Production: Signature Design Group, Inc.
Art Director: Tom Cochrane
Project Director: Carolea Williams

Introduction

The Advantage Math Series for grades 3–8 offers instruction and practice for key skills in each math strand recommended by the National Council for Teachers of Mathematics (NCTM), including

- numeration and number theory
- operations
- geometry
- measurement
- patterns, functions, and algebra
- data analysis and probability
- problem solving

Take a look at all the advantages this math series offers . . .

Strong Skill Instruction

- The **teaching component** at the top of the activity pages provides the support students need to work through the book independently.

- Plenty of **skill practice** pages will ensure students master essential math computation skills they need to increase their math fluency.

- A **problem-solving strand** is woven within skill practice pages to offer students an opportunity to practice critical thinking skills.

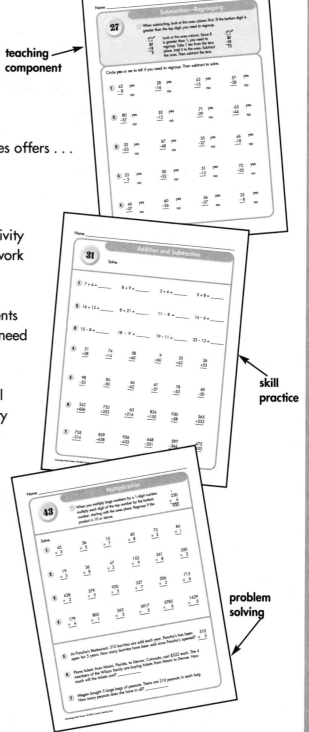

teaching component

skill practice

problem solving

Introduction

- **Mixed-practice pages** include a variety of math concepts on one workbook page. This challenges students to think through each problem rather than rely on a predictable format.

Assessment

- The "Take a Test Drive" pages provide practice using a **test-taking** format such as those included in national standardized and proficiency tests.

- The **tracking sheet** provides a place to record the number of right answers scored on each activity page. Use this as a motivational tool for students to strive for 100% accuracy.

Answer Key

- Answers for each page are provided at the back of the books to make **checking answers quick and easy.**

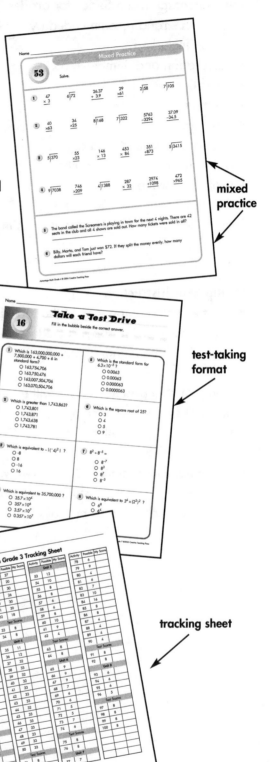

mixed practice

test-taking format

tracking sheet

Place Value

1

	Billions			Millions			Thousands			Hundreds					
	Hundreds	Tens	Ones	Hundreds	Tens	Ones	Hundreds	Tens	Ones	Hundreds	Tens	Ones	Tenths	Hundredths	Thousandths
	10^{11}	10^{10}	10^9	10^8	10^7	10^6	10^5	10^4	10^3	10^2	10^1	10^0	10^{-1}	10^{-2}	10^{-3}
	1	8	4,	5	2	7,	3	6	9,	1	4	6			
											1.	8	3	5	

⭐ **Standard form:**   184,527,369,146
                       1.835

Short word form: 184 billion, 527 million, 369 thousand, 1 hundred, 46
                           1 and 835 thousandths

Expanded form: $1\times10^{11}+8\times10^{10}+4\times10^9+5\times10^8+2\times10^7+7\times10^6+3\times10^5+6\times10^4+9\times10^3+$
                        $1\times10^2+4\times10^1+6\times10^0$
                        $1\times10^0+8\times10^{-1}+3\times10^{-2}+5\times10^{-3}$

Write each number in standard form.

1   105 thousand, 9 hundred, 32 _____

2   six hundred forty-two billion, three hundred thousand _____

3   150,000,000 + 8,000,000 + 6,000 + 5 _____

4   437 billion, 3 million _____

5   fifty-three million, two hundred five thousand _____

6   797,000,000,000 + 619,000,000 +132 _____

7   eleven million, six thousand, seven _____

8   84 billion, 37 thousand, 8 hundred _____

Write the value of the 4 in each number.

9   234,789 _____   747,365,207 _____

10   67,827,412 _____   436,823,706,001 _____

11   43,625,103 _____   693,584,321,787 _____

12   154,600,327,000 _____   270,386,431 _____

Name _____

Comparing and Ordering Integers

2 ⭐ Integers are all the counting numbers (1, 2, 3, 4 . . .), their opposites
(⁻1, ⁻2, ⁻3, ⁻4 . . .) and zero (0).

Comparing numbers on a number line will help you tell if one is greater
than or less than another.

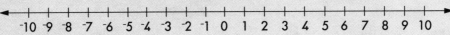

⁻10 ⁻9 ⁻8 ⁻7 ⁻6 ⁻5 ⁻4 ⁻3 ⁻2 ⁻1 0 1 2 3 4 5 6 7 8 9 10

The opposites of the counting   Zero is   The counting numbers are the
numbers are the negative integers.   the center.   positive integers.

If a number is to the left of a number on the number line, it is less than the other.
If it is to the right, it is greater than the other.

Complete the number sentence.

1 1,257 ◯ 1,572   9,800 ◯ 7,327

2 65,382 ◯ 73,721   10,121 ◯ 10,114

3 827,561 ◯ 622,871   1,786,521 ◯ 1,973,804

4 167,256,121 ◯ 17,658,910   986,420 ◯ 73,987,969

5 38,176,200 ◯ 24,298,783   15,627,984,120 ◯ 15,631,256

Order the numbers from least to greatest.

6 987 897 879 978 _____

7 83,586 58,683 68,638 _____

8 107,432,014 423,413,201 407,210,740 _____

9 8,400,327,937 8,437,293,310 8,400,325,831 _____

10 73,627,003 73,603,270 73,302,670 _____

Solve.

11 The attendance for the Yankees game was 37,284 on Tuesday, 37,842 on
Wednesday, and 34,873 on Thursday. Which day had the lowest attendance?

12 Which day had the highest?

Advantage Math Grade 7 © 2005 Creative Teaching Press

Name _____

Absolute Value

3

⭐ The absolute value of an integer is equal to its distance from 0 on a number line.

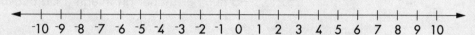

-10 -9 -8 -7 -6 -5 -4 -3 -2 -1 0 1 2 3 4 5 6 7 8 9 10

The absolute value of 3 is 3. The absolute value of ⁻3 is also 3.

$|3| = 3$ because 3 is three units to the right of 0.

$|⁻3| = 3$ because ⁻3 is three units to the left of 0.

Keep in mind that the absolute value bars work differently from parentheses.

$-(⁻3) = ⁺3$. However, $-|⁻3| = ⁻3$. Here's why:

The absolute value of ⁻3 is 3. The negative of an absolute value is a negative number. Try these:

Simplify $|⁻7|$.

$|⁻7| = |7|$

Simplify $-|(⁻2)^2|$.

$-|(⁻2)^2| = -|4| = -4$

Simplify $|0-6|$.

$|0-6| = |-6| = 6$

Simplify $|2+3(-4)|$.

$|2+3(-4)| = |2-12| = |-10| = 10$

Simplify.

1 $|⁻5| =$ _____ $|0-3| =$ _____ $|7| =$ _____

2 $|3-6| =$ _____ $|10-5| =$ _____ $|157-37| =$ _____

3 $|0(⁻2)| =$ _____ $|6(3)| =$ _____ $|⁻2(⁻6)| =$ _____

4 $-|⁻3| =$ _____ $-|7| =$ _____ $-|6(⁻2)| =$ _____

5 $-|(⁻5)^2| =$ _____ $|(⁻8)^3| =$ _____ $-|(⁻6)^4| =$ _____

6 $-|⁻2|^2 =$ _____ $-|4|^3 =$ _____ $(-|3|)^3 =$ _____

Name _____

Scientific Notation

4

 Scientific notation is used in place of very large or very small numbers. Numbers in scientific notation are written as the product of two factors—a number (either an integer or a decimal) and a power of 10.

$$1{,}500{,}000 = 1.5 \times 10^6$$

The **number** has only one digit to the left of the decimal point.

The **power of 10** indicates how many places the decimal point was moved.

Write these numbers using scientific notation.

1 700 = _____   650 = _____   1,300 = _____

2 137,000 = _____   98,700 = _____   1,400,000 = _____

3 8,000,000 = _____   10,000,000 = _____   37,600,000 = _____

4 9,875,000 = _____   68,000,000,000 = _____   25,930,000,000 = _____

Write these numbers in standard form.

5 $1.5 \times 10^2 =$ _____ $8 \times 10^4 =$ _____ $6.3 \times 10^1 =$ _____

6 $1.6 \times 10^3 =$ _____ $1.78 \times 10^5 =$ _____ $1.04 \times 10^2 =$ _____

7 $7.69 \times 10^7 =$ _____ $2.2 \times 10^9 =$ _____ $1.4 \times 10^6 =$ _____

8 $4.78 \times 10^4 =$ _____ $2.76 \times 10^5 =$ _____ $1.87 \times 10^8 =$ _____

Advantage Math Grade 7 © 2005 Creative Teaching Press

Scientific Notation

5 ⭐ The decimal number 0.000046 written in scientific notation would be 4.6×10^{-5} because the decimal point was moved five places to the right to form the number 4.6.

$$0.00345 = 3.45 \times 10^{-3}$$

The <u>number</u> has only one digit to the left of the decimal point.

The <u>power of ten</u> indicates how many places the decimal point was moved.

Write these numbers using scientific notation.

1 $0.017 =$ _____ $0.00462 =$ _____ $0.00874 =$ _____

2 $0.00003 =$ _____ $0.000078 =$ _____ $0.8376 =$ _____

3 $0.0000042 =$ _____ $0.001057 =$ _____ $0.0003608 =$ _____

4 $0.00000008 =$ _____ $0.0006403 =$ _____ $0.0000000428 =$ _____

Write these numbers in standard form.

5 $7.7 \times 10^{-1} =$ _____ $4.3 \times 10^{-3} =$ _____ $5.92 \times 10^{-2} =$ _____

6 $1.06 \times 10^{-7} =$ _____ $5.63 \times 10^{-5} =$ _____ $9.47 \times 10^{-3} =$ _____

7 $7.463 \times 10^{-6} =$ _____ $8.7 \times 10^{-11} =$ _____ $3.76 \times 10^{-4} =$ _____

8 $8.64 \times 10^{-8} =$ _____ $5.397 \times 10^{-9} =$ _____ $2.84 \times 10^{-10} =$ _____

Name _____

6

We say that any number raised to the power of 2 is "squared." The perfect squares are squares of whole numbers. Here are the first three perfect squares.

$$1^2 = 1 \times 1 = 1$$

$$2^2 = 2 \times 2 = 4$$

Any perfect square can be explained using a square!

$$3^2 = 3 \times 3 = 9$$

The square root of a number, n, is a number that when multiplied by itself, equals n. Here are the square roots of the perfect squares above.

$$1^2 = 1 \quad \sqrt{1} = 1$$
$$2^2 = 4 \quad \sqrt{4} = 2$$
$$3^2 = 9 \quad \sqrt{9} = 3$$

Solve.

1 $9^2 =$ _____ $7^2 =$ _____ $10^2 =$ _____ $3^2 =$ _____

2 $\sqrt{25} =$ _____ $\sqrt{36} =$ _____ $\sqrt{100} =$ _____ $\sqrt{64} =$ _____

3 $12^2 =$ _____ $8^2 =$ _____ $15^2 =$ _____ $20^2 =$ _____

4 $\sqrt{49} =$ _____ $\sqrt{400} =$ _____ $\sqrt{225} =$ _____ $\sqrt{196} =$ _____

5 $\sqrt{625} =$ _____ $32^2 =$ _____ $\sqrt{2500} =$ _____ $125^2 =$ _____

Exponents

7

⭐ Here are some other facts to remember about exponents:

The cube of a number is that number multiplied by itself three times.

$$2^3 = 2 \times 2 \times 2 = 8$$

To multiply exponential numbers with the same base, add the exponents.

$$2^3 \times 2^2 = 2^{(3+2)} = 2^5$$
$$8 \times 4 = 32$$

To divide exponential numbers with the same base, subtract the exponents.

$$2^3 \div 2^2 = 2^{(3-2)} = 2^1$$
$$8 \div 4 = 2$$

Solve.

1 $2^3 =$ _____ $4^3 =$ _____ $5^2 \div 5^1 =$ _____ $6^2 \times 6^3 =$ _____

2 $8^4 \div 8^1 =$ _____ $12^6 \div 12^5 =$ _____ $4^3 \times 4^2 =$ _____ $1^7 \div 1^5 =$ _____

3 $4^3 \div 4^2 =$ _____ $2^3 \div 2^1 =$ _____ $8^2 \times 8^1 =$ _____ $10^3 \times 10^2 =$ _____

4 $6^2 \div 6^1 =$ _____ $3^5 \times 3^{-2} =$ _____ $10^{-5} \times 10^3 =$ _____ $4^2 \times 4^{-1} =$ _____

5 $5^1 \div 5^1 =$ _____ $8^3 \times 8^{-1} =$ _____ $0^{-3} \times 0^{-4} =$ _____ $7^4 \div 7^3 =$ _____

Exponents

8

⭐ Here are some more facts to remember about exponents:

The product of the same number to different powers is the number to the sum of the powers.

$$2^3 \times 2^4 = 2^{(3+4)} = 2^7$$

Raising a number to a power, and then raising the result to another power is the same as raising the number to the product of the powers.

$$(2^3)^4 = 2^{(3 \times 4)} = 2^{12}$$

The product of two numbers to the same power is the product of the two numbers raised to the power.

$$2^3 \times 5^3 = (2 \times 5)^3 = 10^3$$

Simplify.

1 $3^2 \times 3^2 =$ _____ $5^1 \times 5^3 =$ _____ $10^5 \times 10^{11} =$ _____ $2^1 \times 2^6 \times 2^3 =$ _____

2 $(8^1)^5 =$ _____ $(7^3)^5 =$ _____ $(4^3)^2 =$ _____ $(3^2)^3 =$ _____

3 $(10^5)^2 =$ _____ $(26^5)^3 =$ _____ $4^6 \times 8^6 =$ _____ $6^8 \times 3^8 =$ _____

4 $2^2 \times 2^3 =$ _____ $4^1 \times (4^2)^1 =$ _____ $(3^2)^5 \times (3^3)^2 =$ _____ $1^2 \times (1^5)^6 =$ _____

5 $(2^3)^1 \times 5^3 =$ _____ $3^4 \times (2^2)^2 =$ _____ $(5^4)^2 \times 3^8 =$ _____ $4^{10} \times 8^{10} =$ _____

Fractional and Negative Exponents

9

⭐ We can use the rules of exponents to understand fractional and negative exponents.

Raising a number to the power ½ is the same as finding its square root. The square root of 9 is 3.

$$(3^2)^{\frac{1}{2}} = 3^{(2 \times \frac{1}{2})} = 3^1 = 3$$

Raising a number to the power ⅓ is the same as finding its cube root. The cube root of 8 is 2.

$$(2^3)^{\frac{1}{3}} = 2^{(3 \times \frac{1}{3})} = 2^1 = 2$$

$$8^{\frac{1}{3}} = 2$$

A number with a negative exponent is the same as the reciprocal of the number with a positive exponent.

$$2^3 \times 2^{-3} = 2^{(3 + ^-3)} = 2^0 = 1$$

Therefore, $2^{-3} = \dfrac{1}{2^3}$.

Because $2^3 \times \dfrac{1}{2^3} = \dfrac{2^3}{2^3} = 1$.

Simplify.

1 $(3^4)^{-2} =$ _____ $(6^{-2})^1 =$ _____ $(10^5)^{-2} =$ _____ $3^{-2} \times 4^{-2} =$ _____

2 $7^{-6} \times 7^5 =$ _____ $8^{-3} \times 6^{-3} =$ _____ $(7^{-4})^{-1} =$ _____ $6^{-3} \times 6^{-5} =$ _____

3 $4^{-10} \times (4^5)^{-2} =$ ___ $(3^{-2})^5 \times (3^{-5})^2 =$ ___ $1^3 \times (1^{-5})^4 =$ ___ $6^{-3} \times (6^3)^{-1} =$ ___

4 $(5^2)^{\frac{1}{2}} =$ _____ $(8^3)^{\frac{1}{3}} =$ _____ $(10^2)^{\frac{1}{2}} =$ _____ $(3^3)^{\frac{1}{3}} =$ _____

Equivalent Fractions

10   ⭐ Multiplying or dividing both terms of a fraction by the same number does not change the value of the fraction.

What fraction with a denominator of 8 is equal to ¼ ?

$$\frac{1}{4} = \frac{?}{8}$$

The first step is to determine how many 4s are contained in 8. The answer is 2, so we know that the multiplier for both terms of the fraction is 2.

$$\frac{1}{4} \times \frac{2}{2} = \frac{2}{8}$$

What fraction with a numerator of 12 is equal to ¾ ? Think: What are the factors of 12? Which factor, when multiplied by 3 equals 12?

$$\frac{3}{4} = \frac{12}{?} \qquad \frac{3}{4} \times \frac{4}{4} = \frac{12}{16}$$

Change ⁸⁄₂₀ to tenths. To reduce a fraction, divide each term by a common factor.

$$\frac{8}{20} = \frac{?}{10} \qquad \frac{8}{20} \div \frac{2}{2} = \frac{4}{10}$$

To reduce a fraction to lowest terms, divide both numerator and denominator by their greatest common factor.

$$\frac{8}{20} = \frac{?}{?} \qquad \frac{8}{20} \div \frac{4}{4} = \frac{2}{5}$$

Solve.

1 $\dfrac{2}{3} = \dfrac{?}{6}$ $\dfrac{4}{5} = \dfrac{16}{?}$ $\dfrac{3}{4} = \dfrac{12}{?}$ $\dfrac{5}{8} = \dfrac{?}{16}$

2 $\dfrac{8}{64} = \dfrac{?}{8}$ $\dfrac{11}{16} = \dfrac{33}{?}$ $\dfrac{9}{10} = \dfrac{?}{30}$ $\dfrac{6}{15} = \dfrac{18}{?}$

3 $\dfrac{12}{16} = \dfrac{3}{?}$ $\dfrac{20}{30} = \dfrac{?}{3}$ $\dfrac{8}{12} = \dfrac{?}{3}$ $\dfrac{14}{28} = \dfrac{?}{2}$

Reduce to lowest terms.

4 $\dfrac{2}{18} =$ $\dfrac{4}{24} =$ $\dfrac{3}{15} =$ $\dfrac{7}{14} =$

5 $\dfrac{6}{18} =$ $\dfrac{4}{32} =$ $\dfrac{7}{28} =$ $\dfrac{12}{32} =$

Decimals and Fractions

11

⭐ To change a fraction into a decimal, divide the numerator by the denominator.

$$\frac{1}{2} = 2\overline{)1.0}^{\,0.5}$$

Sometimes changing a fraction to a decimal results in a nonterminating decimal. Round to find the answer.

$$\frac{4}{9} = 9\overline{)4.000}^{\,0.444} = 0.4$$

To change a decimal to a fraction, use a denominator that is a multiple of 10. Reduce to lowest terms.

$$0.24 = \frac{24}{100} \qquad \frac{24}{100} \div \frac{4}{4} = \frac{6}{25}$$

Convert to decimals.

1 $\frac{3}{8} =$ $\frac{4}{5} =$ $\frac{7}{12} =$

2 $\frac{1}{6} =$ $\frac{3}{7} =$ $\frac{14}{15} =$

3 $\frac{9}{10} =$ $\frac{8}{17} =$ $\frac{6}{7} =$

4 $\frac{20}{32} =$ $\frac{3}{13} =$ $\frac{18}{64} =$

Convert to fractions.

5 $0.25 =$ $0.16 =$ $0.08 =$

6 $0.01 =$ $0.85 =$ $0.33 =$

7 $0.64 =$ $0.45 =$ $0.02 =$

8 $0.05 =$ $0.27 =$ $0.98 =$

Name _____

Finding Percents

12

⭐ To change a fraction to a percent, divide the numerator by the denominator. Multiply by 100 and round the answer if necessary. Add the % sign after the answer.

$$\frac{2}{3} = 3\overline{)2.000}^{0.667} = 0.67 = 67\%$$

To change a percent to a fraction, use a denominator of 100 and reduce to lowest terms.

$$25\% = \frac{25}{100} \qquad \frac{25}{100} \div \frac{25}{25} = \frac{1}{4}$$

To change a decimal to a percent, multiply by 100 by moving the decimal point two places to the right, and add the % sign after the answer.

$$0.554 = 55.4\%$$

To change a percent to a decimal, drop the % sign and divide by 100 by moving the decimal point two places to the left.

$$135\% = 1.35$$
$$52.44\% = 0.5244$$

Change these fractions and decimals to percents.

1 $0.23 =$ $\quad \frac{1}{3} =$ $\quad 0.62 =$ $\quad \frac{3}{4} =$

2 $\frac{7}{8} =$ $\quad 1.10 =$ $\quad \frac{13}{16} =$ $\quad 1.35 =$

3 $\frac{8}{32} =$ $\quad 0.56 =$ $\quad \frac{15}{64} =$ $\quad 10.72 =$

4 $0.03 =$ $\quad \frac{7}{12} =$ $\quad 0.006 =$ $\quad \frac{12}{36} =$

5 $0.0037 =$ $\quad \frac{17}{100} =$ $\quad 1.001 =$ $\quad 10.01 =$

Change these percents to decimals and fractions.

Percent	Decimal	Fraction
20%		
80%		
15%		
35%		
42%		

Advantage Math Grade 7 © 2005 Creative Teaching Press

Estimation and Rounding

13

⭐ To estimate a sum, round each addend, and add the rounded numbers.
235 + 330 = ?
Round 235 to 200 and 330 to 300.
200 + 300 = 500.

If one number is rounded up and the other down, a close estimate will result. However, if both are rounded down, see if the amount of rounding is 50 or greater. If it is, add 100 to the estimate. If both are rounded up and the amount of rounding is 50 or greater, subtract 100 from the estimate.

In this case, both numbers are rounded down: 235 is rounded 35 down to 200, and 330 is rounded 30 down to 300. And 35 + 30 = 65, which is greater than 50. So, add 100 to your estimate: 500 + 100 = 600.

Six hundred is a better estimate than 500. The actual answer is 235 + 330 = 565.

Use the same procedure to estimate a difference. But check those problems in which you rounded one number up and the other down. If you rounded the top number up and the amount of rounding is 50 or greater, subtract 100 from your estimate. If you rounded the top number down and the amount of rounding is 50 or greater, add 100 to your estimate.

Round to estimate the sums and differences.

1
$$\begin{array}{r} 237 \\ + 642 \\ \hline \end{array} \qquad \begin{array}{r} 179 \\ + 151 \\ \hline \end{array} \qquad \begin{array}{r} 405 \\ + 690 \\ \hline \end{array} \qquad \begin{array}{r} 865 \\ + 222 \\ \hline \end{array}$$

2
$$\begin{array}{r} 967 \\ - 388 \\ \hline \end{array} \qquad \begin{array}{r} 420 \\ - 117 \\ \hline \end{array} \qquad \begin{array}{r} 586 \\ - 214 \\ \hline \end{array} \qquad \begin{array}{r} 330 \\ - 280 \\ \hline \end{array}$$

3
$$\begin{array}{r} 737 \\ + 868 \\ \hline \end{array} \qquad \begin{array}{r} 222 \\ + 915 \\ \hline \end{array} \qquad \begin{array}{r} 673 \\ + 155 \\ \hline \end{array} \qquad \begin{array}{r} 193 \\ + 912 \\ \hline \end{array}$$

4
$$\begin{array}{r} 467 \\ - 129 \\ \hline \end{array} \qquad \begin{array}{r} 915 \\ - 639 \\ \hline \end{array} \qquad \begin{array}{r} 464 \\ - 183 \\ \hline \end{array} \qquad \begin{array}{r} 827 \\ - 682 \\ \hline \end{array}$$

Mixed Practice

14

Write each number in standard form.

1. 673 billion, 8 million _____

   ten million, five thousand, two _____

2. 175,000,000 + 4,000,000 + 2,000 + 5 _____

   137 million, 987 thousand _____

Complete the number sentences.

3. 627,651 ◯ 432,863   1,876,1125 ◯ 1,739,048

4. 62,671,350,187 ◯ 32,892,783,056   11,221 ◯ 11,212

Simplify.

5. $|^-4| =$ $|1-3| =$ $|8| =$

6. $-|^-4| =$ $-|(^-6)^2| =$ $(-|^-5|)^3 =$

Write these numbers in standard form.

7. $2.68 \times 10^5 =$ $1.9 \times 10^8 =$ $6.73 \times 10^4 =$

8. $6.5 \times 10^4 =$ $9.47 \times 10^{-3} =$ $3.647 \times 10^{-6} =$

Advantage Math Grade 7 © 2005 Creative Teaching Press

Mixed Practice

15

Solve.

1 $\sqrt{25} =$ $8^2 =$ $\sqrt{196} =$ $32^2 =$

2 $3^2 \div 3^1 =$ $2^3 \times 2^2 =$ $0^5 \times 0^1 =$ $10^5 \times 10^2 =$

Simplify.

3 $(2^2)^3 =$ $3^3 \times 2^3 =$ $(8^2)^4 \times (8^3)^3 =$ $2^2 \times (2^5)^6 =$

4 $7^{-5} \times 7^6 =$ $(2^{-2})^5 \times (2^{-5})^2 =$ $(7^3)^{\frac{1}{3}} =$ $(10^2)^{\frac{1}{2}} =$

Reduce to lowest terms.

5 $\dfrac{7}{8} =$ $\dfrac{28}{84} =$ $\dfrac{56}{112} =$ $\dfrac{26}{32} =$

Convert to decimals or fractions.

6 $\dfrac{5}{6} =$ $\dfrac{3}{9} =$ $\dfrac{6}{8} =$ $\dfrac{2}{7} =$

7 $0.85 =$ $0.98 =$ $0.45 =$ $0.64 =$

Change to percents.

8 $0.004 =$ $\dfrac{7}{12} =$ $1.60 =$ $0.163 =$

9 $\dfrac{8}{32} =$ $0.27 =$ $0.046 =$ $0.31 =$

Name _____

Take a Test Drive

16

Fill in the bubble beside the correct answer.

---

1 Which is 163,000,000,000 + 7,500,000 + 4,700 + 6 in standard form?
- ○ 163,754,706
- ○ 163,750,476
- ○ 163,007,504,706
- ○ 163,070,504,706

5 Which is the standard form for 6.3×10^{-5}?
- ○ 0.0063
- ○ 0.00063
- ○ 0.000063
- ○ 0.0000063

---

2 Which is greater than 1,743,863?
- ○ 1,743,801
- ○ 1,743,871
- ○ 1,743,638
- ○ 1,743,781

6 Which is the square root of 25?
- ○ 3
- ○ 4
- ○ 5
- ○ 9

---

3 Which is equivalent to $-|(^-4)^2|$?
- ○ $^-8$
- ○ 8
- ○ $^-16$
- ○ 16

7 $8^2 \div 8^{-5} =$
- ○ 8^{-7}
- ○ 8^3
- ○ 8^7
- ○ 8^{-3}

---

4 Which is equivalent to 35,700,000?
- ○ 35.7×10^4
- ○ 357×10^6
- ○ 3.57×10^7
- ○ 0.357×10^7

8 Which is equivalent to $3^4 \times (2^2)^2$?
- ○ 6^8
- ○ 6^4
- ○ 5^4
- ○ 5^8

Advantage Math Grade 7 © 2005 Creative Teaching Press

Name _____

1 Which is equivalent to $3^{-10} \times (3^5)^{-2}$?

- ○ 6^{-20}
- ○ 9^{-20}
- ○ 6^{-10}
- ○ 9^{-10}

2 Which number completes the equation $\dfrac{12}{36} = \dfrac{24}{?}$?

- ○ 72
- ○ 48
- ○ 108
- ○ 24

3 Which equals $\dfrac{3}{8}$?

- ○ $\dfrac{27}{64}$
- ○ $\dfrac{27}{36}$
- ○ $\dfrac{27}{72}$
- ○ $\dfrac{27}{24}$

4 Which is equivalent to $\dfrac{4}{5}$?

- ○ 0.50
- ○ 0.80
- ○ 1.25
- ○ 1.50

5 Which is equivalent to 85%?

- ○ $\dfrac{5}{8}$
- ○ $\dfrac{10}{85}$
- ○ $\dfrac{5}{85}$
- ○ $\dfrac{17}{20}$

6 Which is equivalent to 6.003?

- ○ 600.3%
- ○ 60.03%
- ○ 0.06003%
- ○ 6003%

7 Which is equivalent to $\dfrac{7}{8}$?

- ○ 0.0875%
- ○ 8.75%
- ○ 0.875%
- ○ 87.5%

8 Which is the best estimate of the sum 387 + 223?

- ○ 800
- ○ 700
- ○ 600
- ○ 500

Addition

18

⭐ The commutative, associative, distributive, and identity properties of addition apply to all numbers.

Commutative Property: Two numbers can be added in either order.

$$4 + 2 = 2 + 4 \qquad\qquad 3.02 + 0.5 = 0.5 + 3.02$$

Associative Property: When three or more numbers are added, the sum is the same regardless of how they are grouped.

$$(2 + 3) + 4 = 2 + (3 + 4) = (2 + 4) + 3 \qquad (\frac{1}{2} + \frac{1}{3}) + \frac{3}{4} = \frac{1}{2} + (\frac{1}{3} + \frac{3}{4}) = (\frac{1}{2} + \frac{3}{4}) + \frac{1}{3}$$

Distributive Property: The sum of two numbers times a third number is equal to the sum of each addend times the third number.

$$5(6 + 2) = (5 \times 6) + (5 \times 2) \qquad 3(6^2 + 3^2) = (3 \times 6^2) + (3 \times 3^2)$$

Identity Property: The sum of any number and zero is the original number.

$$5 + 0 = 5 \qquad\qquad {}^-4 + 0 = {}^-4$$

Which property of addition is shown?

1 $6.3 + 8.9 = 8.9 + 6.3$

2 $(0.04 + 1.6) + 10.7 = (10.7 + 0.04) + 1$

3 $18 + 27 + 4 = 4 + 27 + 18$

4 $(18\% + 3\%) + 16\% = (3\% + 18\%) + 16\%$

5 $0.00675 + 0 = 0.00675$

6 $\$10.76 + \$5.83 = \$5.83 + 10.76$

7 $118 + 0 = 0$

8 $6^4 + 8^5 = 8^5 + 6^4$

9 $\frac{1}{3}(\frac{1}{8} + \frac{1}{7}) = (\frac{1}{3} \times \frac{1}{8}) + (\frac{1}{3} \times \frac{1}{7})$

10 $10 + 52 + 61 = 52 + 10 + 61$

Advantage Math Grade 7 © 2005 Creative Teaching Press

Adding Whole Numbers

19

⭐ 1 + 4 + 7 = 12. Regroup 12 hundreds as 1 thousand and 2 hundreds.

1 + 6 + 3 = 10. Regroup 10 tens as 1 hundred.

1 + 9 = 10. Regroup 10 ones as 1 ten.

```
  1 1 1
  3,461
+ 4,739
───────
  8,200
```

Add.

1

```
  673        987       3,714      8,972
+  95      + 136      +  867      +   38
```

2

```
6,923        476       8,361        317
+ 1,846     +  78     + 6,439     + 309
```

3

```
1,009      2,708      9,964      1,509
+ 1,597    +  949     +   99     +  890
```

4

```
7,430      4,069      2,495        725
+   360    +  765     +  682     +  349
```

5

```
7,028      4,381      3,768      2,172
+ 3,697    +   86     + 1,357    +  909
```

Adding Whole Numbers

20

Add.

1

$$40,706 \\ + 997$$

$$98,437 \\ + 1,867$$

$$203,189 \\ + 17,643$$

$$987,942 \\ + 638,979$$

2

$$7,356,220 \\ + 15,899$$

$$1,637,923 \\ + 987,569$$

$$4,637,918 \\ + 8,765$$

$$9,678,031 \\ + 18,782$$

3

$$8,137,928 \\ + 4,065,908$$

$$1,653,096 \\ +1,497,961$$

$$7,791,609 \\ + 6,110,987$$

$$5,087,027 \\ + 873,994$$

Solve.

4 If 40,624 tickets to the golf tournament were sold for Friday and 39,879 were sold for Saturday, how many tickets were sold in all?

5 A manufacturing company can produce 1,783,252 cartons of juice a year. If it builds a new plant that can produce 3,246,982 cartons a year, how many cartons of juice can the company produce?

Advantage Math Grade 7 © 2005 Creative Teaching Press

Adding Decimals

21

⭐ Adding numbers with decimals is the same as adding whole numbers. Just make sure to line up the decimal points. Add zeroes to make numbers the same length.

```
  11 11
  34.6670
   3.5000
+244.1875
─────────
 282.3545
```

Add.

1

78.16	136.14	627.143	3.089
9.23	8.63	10.098	6.193
+ 0.97	+ 0.98	+ 1.56	+ 0.009

2

57.783	0.097	0.06731	0.00001
184.02	153.56	0.58932	0.00136
+ 25.104	+ 1.732	+ 0.41437	+ 0.00098

3

1,567.789	0.5647	4,378.5	3.910
134.103	0.93	123.6	4.12
+ 10.12	+ 0.0001	+ 27.51	+ 0.066

4

0.0047	15,676.2	890,123.52	0.867
138.132	10,431.8	24,905.83	0.057
+ 1.964	+ 621.4	+ 0.57	+ 0.507

5

0.01789	0.0637	926	1,714.932
0.03602	17.9651	18.10	0.898
+ 0.02987	+ 182.57	+ 1.657	+ 17.043

Name _____

Adding Decimals

22

Add.

1

0.178	0.00984	763.14	473.972
283.14	0.00638	27.002	110.928
+ 6.982	+ 0.00200	+ 1.7	+ 321.579

2

0.0376	1,624	7.910	0.768
91.7615	257.89	4.89	0.750
+ 158.72	+ 0.013	+ 0.079	+ 0.705

3

1,789,423.16	0.01897	4.793	0.6574
17,987.19	0.06032	1,867.18	0.39
+ 12,094.03	+ 0.08792	+ 32.2876	+ 0.0007

Solve.

4 Raphael wanted to see how much weight the bridge he built out of toothpicks would hold. First, he placed a box that weighed 18.062 ounces on the bridge. Then, he placed another box that weighed 9.643 ounces. When he placed the next box weighing 10.473 ounces, the bridge collapsed. How much weight did Raphael put on the bridge?

5 Sarah made 4 purchases. The cost of each of purchases was $79.84, $12.98, $39.95, and $0.95. What was the total of her purchases?

Advantage Math Grade 7 © 2005 Creative Teaching Press

Adding Integers

23

⭐ The sum of two negative integers is a negative integer.
$^-7+^-4=^-11$

The sum of two positive integers is a positive integer.
$^+7+^+4=^+11$

To add a positive and a negative integer follow these steps.
$^+7+^-4=?$

- Find the absolute value of each integer.
  $|7|+|4|=?$
- Subtract the two numbers you get from Step 1.
  $7-4=3$
- The result takes the sign of the integer with the greatest absolute value.
  $^+7+^-4=^+3$

The sum of an integer and its opposite is 0.
$^+7+^-7=0$

Add.

1 $^+12+^+8=$ $^+32+^+16=$ $^+16+^-3=$ $^+22+^+6=$

2 $^+15+^-12=$ $^+64+^-15=$ $^+29+^-29=$ $^+17+^-12=$

3 $^+167+^+13=$ $^+268+^-68=$ $^+150+^-30=$ $^+87+^-87=$

4 $^+321+^-20=$ $^+787+^+13=$ $^+652+^-37=$ $^+980+^-25=$

5 $^+37.5+^-30.5=$ $^+18.25+^+20.75=$ $^+1.76+^-0.63=$ $^+0.063+^-0.063=$

Name _____

Addition Practice

24

Add.

1
```
   879          1,437          736          7,928
 + 631        +  687        +  59        +   83
```

2
```
  6,873         9,736        83,479       6,473,819
+ 5,371       + 8,207      +  6,187     +     6,578
```

3
```
  7,961,396      9,381,837      7,365.86       0.07678
+ 1,110,687    + 6,046,098    +  213.47     + 0.00996
```

4
```
    0.079         0.000009       8.903        3,785.4
  183.26          0.014026       1.936         321.73
+   7.321       + 0.005720     + 0.008       +  18.0
```

5
```
    0.781         0.4756         1.079         0.00489
  832.14          0.93           8.49          0.00013
+   9.826       + 0.0007       + 0.97        + 0.00064
```

Advantage Math Grade 7 © 2005 Creative Teaching Press

Name _____

25

Add.

1 $^+16 + {}^+8 =$ $^+64 + {}^-13 =$ $^+18.5 + {}^+20.5 =$ $^+38 + {}^-8 =$

2 $^+16.25 + {}^+20.75 =$ $^+13.60 + {}^-2.30 =$ $^+890 + {}^-25 =$ $^+1.68 + {}^-1.68 =$

3 $^+\dfrac{1}{4} + {}^+\dfrac{1}{2} =$ $^+\dfrac{2}{3} + {}^-\dfrac{1}{3} =$ $^+\dfrac{7}{8} + {}^+\dfrac{1}{8} =$ $^+\dfrac{5}{6} + {}^-\dfrac{3}{16} =$

4 $^+0.075 + {}^+0.062 =$ $^+0.032 + {}^-0.002 =$ $^+0.56 + {}^+1.44 =$ $^+0.75 + {}^-0.37 =$

Solve.

5 A large corporation employs 216,641 people in the United States; 3,624 people in Germany; and 15,783 people in Mexico. How many total employees are there?

6 Stephen measured the water in his rain gauge each day for four days. On the first day, he collected 0.085 inches of water. On the second day, he collected 0.035 inches. There was no rain on the third day. On the fourth day, he collected 0.25 inches. How many inches of rain fell over the four days?

7 Susan bought the following at the bookstore: a pen for $2.39, two notebooks for $1.75, and three book covers for $0.96. How much did she spend?

Subtracting Whole Numbers

26

⭐ Rename 6 thousands as 5 thousands and 10 hundreds. Add them to the hundreds.

Rename 8 tens as 7 tens and 10 ones. Add them to the ones.

$$
\begin{array}{r}
{}^{5}\ {}^{14}\ {}^{7}\ {}^{12} \\
6,482 \\
-\ 3,535 \\
\hline
2,947
\end{array}
$$

Subtract.

1

743	326	917	222
− 176	− 193	− 384	− 173

2

1,893	6,321	2,983	4,629
− 863	− 598	− 932	− 849

3

2,893	5,227	8,040	7,941
− 1,576	− 3,988	− 1,635	− 2,906

4

9,010	7,555	3,249	8,493
− 4,364	− 3,666	− 1,249	− 6,009

5

8,979	2,736	4,381	2,573
− 3,897	− 2,697	− 4,090	− 1,697

Advantage Math Grade 7 © 2005 Creative Teaching Press

Subtracting Whole Numbers

27

Subtract.

1
| 18,637 | 98,761 | 67,504 | 32,222 |
| – 6,598 | – 2,197 | – 7,946 | – 1,999 |

2
| 62,731 | 10,020 | 253,832 | 767,932 |
| – 58,176 | – 8,943 | – 79,983 | – 18,408 |

3
| 8,310,402 | 1,005,381 | 4,787,009 | 5,160,980 |
| – 176,185 | – 736,956 | – 2,037,984 | – 2,971,891 |

Solve.

4 The cook at the stadium prepared 17,620 hot dogs to sell to the fans. There were 1,876 hot dogs left at the end of the game. How many hot dogs were purchased by fans?

5 There are 554,636 people who live in Denver, Colorado. There are 8,008,278 people living in New York City. What is the difference in population between the two cities?

Name _____

Subtracting Decimals

28

Subtracting numbers with decimals is the same as subtracting whole numbers. Just make sure to line up the decimal points. Add 0s to make numbers the same length.

Rename 4 thousands as
3 thousands and 10 hundreds.
Add them to the hundreds.

$$
\begin{array}{r}
\overset{3\ 15}{4\,5.683} \\
-\ \ 6.070 \\
\hline
39.613
\end{array}
$$

Subtract.

1)
$$
\begin{array}{r}
12.98 \\
-\ 4.06 \\
\hline
\end{array}
\qquad
\begin{array}{r}
93.64 \\
-\ 29.99 \\
\hline
\end{array}
\qquad
\begin{array}{r}
37.64 \\
-\ 29.99 \\
\hline
\end{array}
\qquad
\begin{array}{r}
324.63 \\
-\ 97.58 \\
\hline
\end{array}
$$

2)
$$
\begin{array}{r}
87.037 \\
-\ 1.569 \\
\hline
\end{array}
\qquad
\begin{array}{r}
43.569 \\
-\ 9.072 \\
\hline
\end{array}
\qquad
\begin{array}{r}
79.683 \\
-\ 5.29 \\
\hline
\end{array}
\qquad
\begin{array}{r}
27.910 \\
-\ 18.664 \\
\hline
\end{array}
$$

3)
$$
\begin{array}{r}
127.59 \\
-\ 39.50 \\
\hline
\end{array}
\qquad
\begin{array}{r}
767.518 \\
-\ 15.069 \\
\hline
\end{array}
\qquad
\begin{array}{r}
1,248.632 \\
-\ 16.988 \\
\hline
\end{array}
\qquad
\begin{array}{r}
8,673.593 \\
-\ 6,082.094 \\
\hline
\end{array}
$$

Solve.

4) Jonathan purchased a new table saw for $1,827.63. He gave the cashier $1,900.00. How much change should he get back?

5) A large truck pulled off into a weigh station. The truck weighed 50,873.563 pounds. If the legal weight for trucks is 55,000 pounds, how much more weight could this truck carry?

Advantage Math Grade 7 © 2005 Creative Teaching Press

Name _____

Subtracting Integers

29

$^-7-^-4$ is the same as $-(7-4)$. $-(7-4)=^-3$

$^+7-^+4$ is the same as $7-4$. $7-4=3$

$^+7-^-4$ is the same as $7+4$. $7+4=11$

Subtract.

1. $^+3-^+6 =$ $^-3-^-6 =$ $^-10-^-6 =$ $^+12-^-6 =$

2. $^-32-^+8 =$ $^+16-^-6 =$ $^+24-^+12 =$ $^-15-^-15 =$

3. $^-37-0 =$ $0-^-37 =$ $0-^+37 =$ $^+82-^-12 =$

4. $^+4.5-^+1.6 =$ $^-8.3-^-4.8 =$ $^+16.7-^-14.3 =$ $^+32.6-^+24.2 =$

5. $^+\dfrac{1}{2}-^+\dfrac{1}{4} =$ $^-\dfrac{3}{4}-^-\dfrac{1}{4} =$ $^+\dfrac{3}{4}-^+\dfrac{1}{3} =$ $^-\dfrac{3}{3}-^-\dfrac{1}{3} =$

Addition and Subtraction Equations

30

⭐ Remember that whatever you do to one side of an equation, you must also do to the other side to keep it in balance.

$$x + 15 = 30$$
$$x + 15 - 15 = 30 - 15$$
$$x = 15$$
$$y - 6 = 3$$
$$y - 6 + 6 = 3 + 6$$
$$y = 9$$

Check by substituting the value 9 for y in the original equation.

$$y - 6 = 3$$
$$(9) - 6 = 3$$
$$3 = 3 \text{ correct!}$$

Solve.

1 $x + 6 = 10$ $x + 32 = 64$ $x + 72 = 100$
$x =$ $x =$ $x =$

2 $x - 15 = 30$ $x - 14 = 45$ $x - 52 = 75$
$x =$ $x =$ $x =$

3 $x + 126 = 150$ $x - 37 = 162$ $x + 487 = 625$
$x =$ $x =$ $x =$

4 $x - 1,624 = 132$ $x + 7,897 = 9,800$ $x - 983 = 172$
$x =$ $x =$ $x =$

Advantage Math Grade 7 © 2005 Creative Teaching Press

Subtraction Practice

31

Subtract.

1

2,316	9,634	7,100	4,197
− 859	− 489	− 6,436	− 2,690

2

76,189	21,064	5,013,817	6,150,097
− 9,172	− 9,438	− 367,659	− 2,179,981

3

64.37	89.98	91.207	79.583
− 47.64	− 10.98	− 18.368	− 67.666

4

1,657.289	677.815	6,837.953	7,362.025
− 283.509	− 19.908	− 6,028.904	− 5,987.08

5 $^-89-^-32 =$ $^+64-^+12 =$ $^-18-^-18 =$ $^+23-^+17 =$

Subtraction Practice

32

Solve.

1 $^+8.3 - {}^+6.5 =$ $^-9.8 - {}^-8.4 =$ $^+15.8 - {}^-12.3 =$ $^+44.67 - {}^+18.73 =$

2 $x + 17 = 68$ $x + 12 = 23$ $x + 72 = 98$ $x + 1,060 = 2,100$

3 $x - 57 = 43$ $x - 17 = 163$ $x - \dfrac{1}{4} = \dfrac{1}{2}$ $x - \dfrac{3}{8} = \dfrac{2}{8}$

Solve.

4   Latisha guessed there were 1,627,182 jelly beans in the huge glass jar. Bonnie guessed there were 1,047,832. What is the difference between their guesses?

5   Chris is collecting baseball cards this year. Currently, he has 173. There are 536 in the complete collection. How many more cards does he need to complete his collection?

6   To complete a marathon, a person has to run 26.2 miles. If Jill has run 7.5 miles, how many more miles does she have to go to complete the marathon?

Name _____

Factors and Multiples

33

⭐ Numbers that are multiplied together to create a new number are called the factors of that new number. For example, $5 \times 4 = 20$. Five and 4 are factors of 20. Two and 10 are also factors of 20 because $2 \times 10 = 20$. One and 20 are also factors, because $1 \times 20 = 20$.

All even numbers have a factor of 2.
All numbers ending in 5 have a factor of 5.
All numbers greater than 0 that end with 0 have factors of 2 and 5.

Here are the factors of some numbers:

6: 1, 2, 3, 6   14: 1, 2, 7, 14   21: 1, 3, 7, 21
12: 1, 2, 3, 4, 6, 12   15: 1, 3, 5, 15

The greatest common factor (GCF) of 6 and 12 is 6.
The GCF of 14 and 21 is 7.
The GCF of 6 and 15 is 3.

A **multiple** of a number is that number times another number. For example, the first five multiples of various numbers are:

2: 2, 4, 6, 8, 10   6: 6, 12, 18, 24, 30   12: 12, 24, 36, 48, 60

To find the least common multiple (LCM) of two numbers:
- Find the GCF of the numbers.
- Multiply the numbers together. Divide the product of the numbers by the GCF.

What is LCM of 15 and 12?
- Find the GCF of 15 and 12, which is 3.
- Multiply the numbers and divide by the GCF.
  $15 \times 12 = 180$; $180 \div 3 = 60$

Show the factors of these numbers.

1. 18 _____   45 _____

2. 36 _____   49 _____

List the first five multiples of these numbers.

3. 8 _____   9 _____

4. 11 _____   20 _____

Name _____

34

Show the factors of each number. Circle the greatest common factor (GCF).

1. 24 _____   32 _____

2. 15 _____   27 _____

3. 64 _____   30 _____

4. 50 _____   48 _____

What is the greatest common factor (GCF) of each pair?

5. 8 and 16 _____   49 and 63 _____

6. 24 and 64 _____   25 and 75 _____

7. 15 and 30 _____   27 and 54 _____

8. 21 and 49 _____   36 and 54 _____

Prime and Composite Numbers

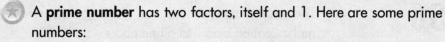

35

⭐ A **prime number** has two factors, itself and 1. Here are some prime numbers:

$$3 = 3 \times 1 \qquad 17 = 17 \times 1 \qquad 43 = 43 \times 1$$

A composite number has more than two factors. Here are some composite numbers:

$49 = 1 \times 49$ and 7×7 $\qquad 64 = 1 \times 64$, 2×32, 4×16, and 8×8

Since all even numbers have a factor of 2, every even number greater than 2 is a composite number.

All numbers that end in 5 have a factor of 5. Therefore all numbers that end with 5 and are greater than 5 are composite numbers.

The numbers 0 and 1 are neither prime nor composite. However, all numbers that end in 0 and are greater than 0 have factors of 2 and 5. Therefore, all numbers greater than 0 that end in 0 are composite numbers.

Tell whether each number is prime or composite, and show the factors for each number.

1 7 _____ 15 _____ 8 _____

2 32 _____ 37 _____ 18 _____

3 56 _____ 70 _____ 24 _____

4 111 _____ 81 _____ 35 _____

5 83 _____ 49 _____ 19 _____

6 47 _____ 96 _____ 51 _____

Multiplication

36

⭐ The commutative, associative, distributive, and identity properties of multiplication apply to all numbers.

Commutative Property: Two numbers can be multiplied in either order.

$4 \times 2 = 2 \times 4$ $\qquad\qquad$ $3.02 \times 0.5 = 0.5 \times 3.02$

Associative Property: When three or more numbers are multiplied, the product is the same regardless of how they are grouped.

$2 \times (3 \times 4) = (2 \times 4) \times 3$ $\qquad$ $(\frac{1}{2} \times \frac{1}{3}) \times \frac{3}{4} = \frac{1}{2} \times (\frac{1}{3} \times \frac{3}{4}) = (\frac{1}{2} \times \frac{3}{4}) \times \frac{1}{3}$

Distributive Property: The sum of two numbers times a third number is equal to the sum of each addend times the third number.

$5(6 + 2) = (5 \times 6) + (5 \times 2)$ $\qquad$ $3(6^2 + 3^2) = (3 \times 6^2) + (3 \times 3^2)$

Identity Property: The product of any number and one is that number.

$5 \times 1 = 5$ $\qquad$ $^-4 \times {}^+1 = {}^-4$

Which property of multiplication is shown?

1 $\quad 6 \times 8 = 8 \times 6$ $\qquad\qquad$ _____

2 $\quad 12(4 + 3) = (12 \times 4) + (12 \times 3)$ $\qquad$ _____

3 $\quad 3(6 \times 8) = 6(3 \times 8)$ $\qquad\qquad$ _____

4 $\quad \frac{1}{9} \times 1 = \frac{1}{9}$ $\qquad\qquad$ _____

5 $\quad 15\% \times 18\% = 18\% \times 15\%$ $\qquad$ _____

6 $\quad 68.72 \times 1 = 68.72$ $\qquad\qquad$ _____

7 $\quad (15 \times 3)8 = 15(8 \times 3)$ $\qquad\qquad$ _____

8 $\quad 12^2 \times 6^3 = 6^3 \times 12^2$ $\qquad\qquad$ _____

9 $\quad 6(\$1.32 + \$0.75) = (6 \times \$1.32) + (6 \times \$0.75)$ $\qquad$ _____

10 $\quad \frac{1}{9}(\frac{1}{6} \times \frac{1}{5}) = \frac{1}{5}(\frac{1}{9} \times \frac{1}{6})$ $\qquad$ _____

Multiplying Whole Numbers

37

⭐ (8 x 9) + 3 = 75 tens. Regroup 75 tens as 7 hundreds and 5 tens.
8 x 4 = 32 ones. Regroup 32 ones as 3 tens and 2 ones.

```
  7 3
6,294
×    8
─────
50,352
```

Multiply.

1
```
  637        863        392        573
×   5      ×   9      ×   8      ×   6
```

2
```
1,284      6,723      5,103      6,130
×    3     ×    8     ×    9     ×    7
```

3
```
3,214      2,016      9,107      1,987
×    5     ×    4     ×    3     ×    6
```

4
```
7,624      9,843      4,270      5,030
×    5     ×    2     ×    6     ×    9
```

5
```
4,123      2,367      7,109      9,178
×    6     ×    9     ×    5     ×    3
```

6
```
1,062      7,019      8,791      3,105
×    7     ×    8     ×    4     ×    5
```

Multiplying Whole Numbers

38

$$
\begin{array}{r}
5{,}329 \\
\times \quad 43 \\
\hline
15{,}987 \\
+\;213{,}160 \\
\hline
229{,}147
\end{array}
$$

First, multiply by the 3 in the ones place.

Then, multiply by the 4 in the tens place.

Then, add the partial products to find the answer.

Multiply.

1

6,763	1,356	6,922	7,805
× 21	× 17	× 27	× 13

2

1,040	5,500	9,574	5,568
× 34	× 20	× 66	× 87

3

9,806	6,624	4,147	1,603
× 90	× 53	× 98	× 75

4

1,274	4,202	8,326	1,480
× 49	× 60	× 78	× 80

Advantage Math Grade 7 © 2005 Creative Teaching Press

Multiplying Whole Numbers

39

Multiply.

1

201	7,973	1,873	15,176
× 8	× 5	× 13	× 10

2

60,810	263,000	521,432	47,345
× 53	× 15	× 11	× 98

3

98,000	344,020	941,561	68,405
× 32	× 61	× 38	× 88

4

675,455	177,000	252,210	188,889
× 46	× 19	× 76	× 46

Multiplying Decimals

40

⭐ Multiplying numbers with decimals is the same as multiplying whole numbers. Just make sure to line up the decimal points. Add 0s to make numbers the same length.

4.343 x 0.67 = ?

$$
\begin{array}{r}
4.343 \\
\times\ 0.670 \\
\hline
304010 \\
+2605800 \\
\hline
2909810
\end{array}
$$

Add up the number of places after decimal points in the factors. Place the decimal point in the product that many places from the right. Zeroes at the end of the product can be dropped.

4.343 x 0.67 = 2.909810, or 2.90981

Multiply.

1

1.37	0.913	73.1	33.18
× 0.71	× 3.4	× 0.6	× 0.77

2

2.55	18	0.03	46.2
× 1.99	× 1.8	× 0.09	× 37.1

3

0.65	0.061	12.33	0.403
× 5.3	× 0.02	× 0.5	× 10

4

0.065	52.2	0.1515	1.25
× 0.7	× 0.1	× 1.05	× 0.75

Multiplying Integers

41

⭐ A positive number multiplied by a negative number always results in a negative number.

$^+30\times^-6=^-180$

A negative number multiplied by a negative number always results in a positive number.

$^-8\times^-3=^+24$

A positive number multiplied by a positive number always results in a positive number.

$^+12\times^+8=^+96$

Multiply.

1 $^+4\times^+12 =$ $^+12\times^-4 =$ $^-6\times^+8 =$ $^-3\times^-8 =$

2 $^+12\times^-3 =$ $^-11\times^+3 =$ $^-10\times^-15 =$ $^+16\times^-2 =$

3 $^-11\times^-11 =$ $^-20\times^+5 =$ $^+18\times^-9 =$ $^+17\times^-1 =$

4 $^-65\times^+2 =$ $^+6\times^+14 =$ $^+22\times^-4 =$ $^-15\times^-5 =$

5 $^-100\times^-4 =$ $^+39\times^-4 =$ $^-12\times^-6 =$ $^+120\times^-7 =$

6 $^-160\times^+30 =$ $^-180\times^-20 =$ $^+250\times^+5 =$ $^+425\times^-4 =$

Dividing Whole Numbers

42 ⭐ 5652 ÷ 24 = ?

```
        235
    24)5652
        48
        85
        72
       132
       120
        12
```

The remainder is 12: 5,652 ÷ 24 = 235 R12.

Divide.

1   6)3,270   4)9,871   8)1,323   5)9,762

2   12)6,420   32)5,872   12)7,362   86)5,762

3   25)2,550   79)4,424   80)9,050   29)3,016

4   42)1,260   92)5,245   32)2,576   35)1,470

Advantage Math Grade 7 © 2005 Creative Teaching Press

Dividing Whole Numbers

43

⭐ Dividing without leaving a remainder can result in a decimal quotient.

$5652 \div 24 = ?$

$5652 \div 24 = 235.5$

$$
\begin{array}{r}
235.5 \\
24\overline{)5652.0} \\
\underline{48} \\
85 \\
\underline{72} \\
132 \\
\underline{120} \\
120 \\
\underline{120}
\end{array}
$$

Divide.

1　　$5\overline{)52,834}$　　　　$8\overline{)13,064}$　　　　$4\overline{)98,762}$　　　　$12\overline{)36,765}$

2　　$16\overline{)51,836}$　　　$25\overline{)73,210}$　　　$20\overline{)18,975}$　　　$47\overline{)10,998}$

3　　$35\overline{)83,622}$　　　$32\overline{)74,864}$　　　$75\overline{)68,235}$　　　$95\overline{)13,794}$

Solve.

4 The 18 classes at Pearson School collected \$24,830 in a fund-raising drive. On average, how much money was collected by each class?

Name _____

Dividing Decimals

44

⭐ Place the decimal point in the quotient above the decimal point in the dividend.

$$\begin{array}{r} 2.05 \\ 8\overline{)16.40} \\ \underline{16} \\ 040 \\ \underline{40} \\ 0 \end{array}$$

Divide.

1 $9\overline{)36.81}$ $6\overline{)30.42}$ $8\overline{)64.32}$ $4\overline{)500.20}$

2 $5\overline{)125.45}$ $12\overline{)672.24}$ $18\overline{)666.72}$ $42\overline{)546.84}$

3 $14\overline{)171.50}$ $35\overline{)551.25}$ $72\overline{)406.80}$ $16\overline{)341.60}$

Solve.

4 All twelve members of the DiMarco family decided to buy a go-cart and split the cost. If the go-cart costs $790.20, how much will each family member have to spend?

Advantage Math Grade 7 © 2005 Creative Teaching Press

Dividing Decimals

45

If both divisor and dividend are decimals, count the number of decimal places in the divisor. Move the decimal points in both divisor and dividend that many places to the left.
Then place the decimal point in the quotient above the decimal point in the dividend.

$$8.5\overline{)19.55}$$

$$\begin{array}{r} 2.3 \\ 85\overline{)195.5} \\ \underline{170} \\ 255 \\ \underline{255} \\ 0 \end{array}$$

Divide.

1 $5.3\overline{)34.45}$ $7.8\overline{)98.28}$ $3.9\overline{)99.84}$ $6.7\overline{)97.15}$

2 $12.5\overline{)320}$ $18.6\overline{)842.58}$ $11.2\overline{)512.96}$ $25.5\overline{)471.75}$

3 $36.3\overline{)958.32}$ $68.4\overline{)861.84}$ $27.3\overline{)431.34}$ $10.6\overline{)334.96}$

Solve.

4 A garden 39.2 meters wide is being divided into sections 5.6 meters wide. How many sections will there be?

5 Steve is cutting a 190.4 meter rope into pieces 1.4 meters long. How many pieces will he have?

Dividing Integers

46

⭐ When dividing integers, keep these rules in mind:

If both the dividend and divisor are positive, the quotient will be positive.
$^+56 \div ^+7 = ^+8$

If both the dividend and divisor are negative, the quotient will be positive.
$^-56 \div ^-7 = ^+8$

If either the dividend or divisor is negative and the other is positive, the quotient will be negative.
$^+56 \div ^-7 = ^-8$

Divide.

1 $^+32 \div ^+8 =$ $^-72 \div ^+9 =$ $^-64 \div ^-8 =$

2 $^+108 \div ^-7.2 =$ $^-56.61 \div ^-3.7 =$ $^+598.5 \div ^+7 =$

3 $^-658.98 \div ^-52.3 =$ $^-779.16 \div ^+17.2 =$ $^+387.60 \div ^+8.5 =$

4 $^-803.66 \div ^+14.3 =$ $^+559.18 \div ^+38.3 =$ $^-969.44 \div ^-66.4 =$

5 $^+144 \div ^-12 =$ $^+396.9 \div ^-16.2 =$ $^-326.5 \div ^+5 =$

Mixed Practice

47

Solve.

1

9,382	4,080	56,704	22,322
+ 5,176	− 3,561	+ 9,746	− 9,199

2

3,810,204	1,605,080	64.37	234.36
− 671,581	− 981,918	+ 92.63	− 79.85

3

677.518	89.386	3,786.395	4,128.236
+ 15.096	− 2.95	− 8,062.409	+ 761.898

4 $^-13-^-13=$ $^-15-^-8=$ $^+16+^-3=$ $^+7.9-^+6.4=$

5 $^+71-^-13=$ $0-^-48=$ $^+16+^-6=$ $^-63-0=$

Mixed Practice

48

Solve.

1 $x + 12 = 36$ $x + 8.5 = 20$ $y - 13 = 33$ $y - 63 = 27$
   $x =$ $x =$ $y =$ $y =$

2
$$\begin{array}{r} 735 \\ \times\ \ \ 8 \\ \hline \end{array}$$
$$\begin{array}{r} 3{,}267 \\ \times\ \ \ \ 4 \\ \hline \end{array}$$
$$\begin{array}{r} 3{,}050 \\ \times\ \ \ 29 \\ \hline \end{array}$$
$$\begin{array}{r} 48{,}650 \\ \times\ \ \ \ 37 \\ \hline \end{array}$$

3
$$\begin{array}{r} 18.33 \\ \times\ 0.66 \\ \hline \end{array}$$
$$\begin{array}{r} 0.06 \\ \times\ 0.08 \\ \hline \end{array}$$
$$\begin{array}{r} 0.416 \\ \times\ 12 \\ \hline \end{array}$$
$$\begin{array}{r} 1.33 \\ \times\ 0.65 \\ \hline \end{array}$$

4 $^-15 \times {}^+5 =$ $^-17 \times {}^-1 =$ $^-64 \div {}^-8 =$ $^+144 \div {}^-12 =$

5 $29\overline{)1{,}305}$ $64\overline{)8{,}000}$ $56\overline{)13{,}748}$ $95\overline{)26{,}448}$

6 $10.5\overline{)163.8}$ $17.3\overline{)788.88}$ $29.3\overline{)741.29}$ $84.2\overline{)715.70}$

Take a Test Drive

49

Fill in the bubble beside the correct answer.

---

1 Which property of addition is shown?
$(0.03 + 1.6) + 7.5 = 0.03 + (1.6 + 7.5)$

- ○ associative
- ○ commutative
- ○ distributive
- ○ identity

---

5 Annika spent $37.42, $9.15, and $29.99 on footwear. How much did she spend in all?

- ○ $46.56
- ○ $76.56
- ○ $146.56
- ○ $158.91

2
$$3,183$$
$$+\ 1,596$$

- ○ 3,680
- ○ 3,780
- ○ 4,680
- ○ 4,780

---

6 $^{+}28 + {}^{-}16 =$

- ○ −12
- ○ +12
- ○ −44
- ○ +44

---

3 $1,251,080 + 309,207 =$

- ○ 1,560,287
- ○ 1,561,007
- ○ 2,650,287
- ○ 4,343,150

---

7 $\dfrac{^{+}3}{4} + \dfrac{^{-}1}{4} =$

- ○ 1
- ○ $\dfrac{1}{2}$
- ○ $^{-}1$
- ○ $\dfrac{^{-}1}{2}$

---

4 $0.45 + 1.63 + 26.3 =$

- ○ 0.4711
- ○ 4.711
- ○ 28.38
- ○ 283.81

8
$$8,176$$
$$-\ \ \ \ 791$$

- ○ 8,967
- ○ 8,685
- ○ 7,425
- ○ 7,385

Take a Test Drive

50

Fill in the bubble beside the correct answer.

1
   16.35
  − 7.66

- ○ 9.3
- ○ 9.69
- ○ 8.69
- ○ 24.01

---

5 Franco didn't know he had a hole in his pocket. He lost $6.75 on his way to school and another five cents on his way home. Which could be used to figure out how much he lost?

- ○ 6.75 + 0.5
- ○ ⁻6.75 + ⁺0.05
- ○ ⁻6.75 − ⁻0.05
- ○ ⁻6.75 + ⁻0.05

---

2 1,353.04 − 352.003 =

- ○ 1,001.037
- ○ 1,000.037
- ○ 1,001.01
- ○ 1,000.01

---

6 Which is the Least Common Multiple of 7 and 13?

- ○ 36
- ○ 52
- ○ 63
- ○ 91

---

3 ⁻9.2 − ⁻7.1 =

- ○ -2.1
- ○ +2.1
- ○ -16.3
- ○ +16.3

---

7 26)2,587

- ○ 99 R13
- ○ 99.5
- ○ 99.115
- ○ 99.25

---

4 x − 63 = 5
x =

- ○ 61
- ○ 58
- ○ 68
- ○ 56

---

8 Sarah is cutting a board that is 14.4 feet long into smaller lengths for shelves. How many pieces that are 1.2 feet long can she get out of the board?

- ○ 12
- ○ 12.5
- ○ 12.4
- ○ 13

Advantage Math Grade 7 © 2005 Creative Teaching Press

Name _____

51

⭐ To add or subtract fractions with like denominators, just add or subtract the numerators. To simplify, reduce the sum or difference to lowest terms.

$$\frac{3}{5} + \frac{4}{5} = ?$$

$$\frac{3}{5} + \frac{4}{5} = \frac{(3+4)}{5} = \frac{7}{5}$$

$$\frac{7}{5} = 1\frac{2}{5}$$

$$\frac{7}{8} - \frac{5}{8} = ?$$

$$\frac{7}{8} - \frac{5}{8} = \frac{(7-5)}{8} = \frac{2}{8}$$

$$\frac{2}{8} = \frac{1}{4}$$

Add or subtract.

1 $\dfrac{1}{6} + \dfrac{2}{6} =$ $\qquad$ $\dfrac{2}{5} + \dfrac{4}{5} =$ $\qquad$ $\dfrac{1}{8} + \dfrac{6}{8} =$ $\qquad$ $\dfrac{3}{13} + \dfrac{7}{13} =$

2 $\dfrac{8}{9} - \dfrac{4}{9} =$ $\qquad$ $\dfrac{7}{8} - \dfrac{3}{8} =$ $\qquad$ $\dfrac{7}{10} - \dfrac{6}{10} =$ $\qquad$ $\dfrac{15}{32} - \dfrac{8}{32} =$

3 $\dfrac{19}{26} - \dfrac{3}{26} =$ $\qquad$ $\dfrac{18}{35} + \dfrac{6}{35} =$ $\qquad$ $\dfrac{15}{72} - \dfrac{7}{72} =$ $\qquad$ $\dfrac{16}{113} + \dfrac{24}{113} =$

4 $\dfrac{19}{64} + \dfrac{6}{64} =$ $\qquad$ $\dfrac{17}{168} - \dfrac{3}{168} =$ $\qquad$ $\dfrac{16}{19} + \dfrac{2}{19} =$ $\qquad$ $\dfrac{14}{31} - \dfrac{12}{31} =$

Adding and Subtracting Fractions

52

⭐ To add or subtract fractions with unlike denominators, convert to equivalent fractions using the least common multiple (LCM) as the denominator.

$\dfrac{3}{4} + \dfrac{5}{6} = ?$ The LCM of 4 and 6 is 12.

$\dfrac{3}{4} = \dfrac{9}{12}$

$\dfrac{5}{6} = \dfrac{10}{12}$

$\dfrac{9}{12} + \dfrac{10}{12} = \dfrac{19}{12} = 1\dfrac{7}{12}$

$\dfrac{5}{6} - \dfrac{1}{3} = ?$ The LCM of 6 and 3 is 6.

$\dfrac{1}{3} = \dfrac{2}{6}$

$\dfrac{5}{6} - \dfrac{2}{6} = \dfrac{3}{6} = \dfrac{1}{2}$

Add or subtract.

1 $\dfrac{1}{6} + \dfrac{2}{5} =$ $\qquad$ $\dfrac{1}{5} + \dfrac{2}{8} =$ $\qquad$ $\dfrac{1}{4} + \dfrac{1}{3} =$ $\qquad$ $\dfrac{3}{5} + \dfrac{1}{7} =$

2 $\dfrac{1}{3} + \dfrac{1}{6} =$ $\qquad$ $\dfrac{2}{7} + \dfrac{1}{21} =$ $\qquad$ $\dfrac{3}{16} + \dfrac{5}{8} =$ $\qquad$ $\dfrac{6}{7} + \dfrac{1}{2} =$

3 $\dfrac{2}{3} - \dfrac{1}{5} =$ $\qquad$ $\dfrac{6}{7} - \dfrac{1}{2} =$ $\qquad$ $\dfrac{3}{8} - \dfrac{1}{4} =$ $\qquad$ $\dfrac{7}{10} - \dfrac{1}{5} =$

4 $\dfrac{4}{5} - \dfrac{1}{4} =$ $\qquad$ $\dfrac{4}{7} - \dfrac{1}{3} =$ $\qquad$ $\dfrac{7}{12} - \dfrac{1}{8} =$ $\qquad$ $\dfrac{5}{6} - \dfrac{2}{7} =$

 Advantage Math Grade 7 © 2005 Creative Teaching Press

Adding and Subtracting Integer Fractions

53

⭐ To add and subtract integer fractions, make the numerator of any negative fraction negative and the denominator positive.

$$-\frac{3}{4} - (-\frac{1}{3}) = ?$$

$$\frac{-3}{4} - \frac{-1}{3} = \text{The LCM of 4 and 3 is 12.}$$

$$\frac{-3}{4} = \frac{-9}{12}$$

$$\frac{-1}{3} = \frac{-4}{12}$$

$$\frac{-9}{12} - \frac{-4}{12} = \frac{-(9-4)}{12} = \frac{-5}{12}$$

$$-\frac{2}{7} + 1\frac{2}{7} =$$

$$\frac{-2}{7} + 1\frac{2}{7} = \frac{-(2+9)}{7} = \frac{7}{7} = 1$$

Add or subtract.

1 $\quad -\frac{4}{5} - (-\frac{3}{5}) = \qquad\quad \frac{7}{8} - (-\frac{3}{8}) = \qquad\quad -\frac{5}{9} + \frac{2}{9} = \qquad\quad 1\frac{2}{7} - (-\frac{5}{7}) =$

2 $\quad -\frac{2}{3} - (-\frac{1}{4}) = \qquad\quad -\frac{3}{8} - (-\frac{1}{5}) = \qquad\quad \frac{6}{7} - (-\frac{3}{4}) = \qquad\quad -1\frac{7}{8} - (-\frac{4}{5}) =$

3 $\quad \frac{3}{4} + 1\frac{1}{4} = \qquad\quad -\frac{6}{7} + 2\frac{2}{7} = \qquad\quad -\frac{5}{8} + 1\frac{7}{8} = \qquad\quad \frac{3}{16} + 1\frac{7}{16} =$

4 $\quad -\frac{2}{9} + 3\frac{1}{9} = \qquad\quad -\frac{4}{5} + 2\frac{3}{5} = \qquad\quad -\frac{11}{13} + 2\frac{3}{13} = \qquad\quad -\frac{3}{35} + 4\frac{10}{35} =$

Addition and Subtraction of Fractions Practice

54

Add or subtract.

1) $\dfrac{4}{5} + \dfrac{3}{5} =$ $\dfrac{4}{13} + \dfrac{9}{13} =$ $\dfrac{15}{73} - \dfrac{8}{73} =$ $\dfrac{14}{27} - \dfrac{23}{27} =$

2) $\dfrac{27}{39} - \dfrac{16}{39} =$ $\dfrac{17}{32} + \dfrac{9}{32} =$ $\dfrac{115}{326} - \dfrac{72}{326} =$ $\dfrac{64}{92} + \dfrac{82}{92} =$

3) $\dfrac{3}{8} + \dfrac{1}{3} =$ $\dfrac{4}{5} - \dfrac{1}{6} =$ $\dfrac{3}{16} + \dfrac{3}{8} =$ $\dfrac{4}{9} + \dfrac{1}{4} =$

4) $\dfrac{3}{13} + \dfrac{5}{26} =$ $\dfrac{7}{12} - \dfrac{1}{8} =$ $\dfrac{3}{8} - \dfrac{2}{5} =$ $\dfrac{6}{7} - \dfrac{1}{2} =$

5) $\dfrac{7}{9} - \dfrac{1}{2} =$ $\dfrac{7}{8} - \dfrac{3}{4} =$ $\dfrac{3}{7} + \dfrac{1}{2} =$ $\dfrac{2}{3} - \dfrac{1}{5} =$

6) $\dfrac{9}{16} + \dfrac{1}{3} =$ $\dfrac{5}{8} + \dfrac{2}{3} =$ $\dfrac{16}{35} - \dfrac{4}{5} =$ $\dfrac{7}{10} - \dfrac{1}{2} =$

Addition and Subtraction of Fractions Practice

55

Add or subtract.

1 $\dfrac{1}{2} - \dfrac{1}{9} =$ $\dfrac{7}{10} - \dfrac{2}{5} =$ $\dfrac{7}{8} - \dfrac{2}{7} =$ $\dfrac{5}{12} + \dfrac{3}{16} =$

2 $\dfrac{7}{8} - \dfrac{11}{16} =$ $\dfrac{8}{9} + \dfrac{5}{8} =$ $\dfrac{3}{5} - \dfrac{3}{16} =$ $\dfrac{8}{9} + \dfrac{6}{12} =$

3 $\dfrac{3}{10} - \dfrac{3}{15} =$ $\dfrac{11}{12} - \dfrac{1}{4} =$ $\dfrac{1}{6} + \dfrac{1}{8} =$ $\dfrac{5}{6} - \dfrac{1}{9} =$

4 $\dfrac{7}{8} - \left(-\dfrac{3}{8}\right) =$ $-\dfrac{4}{9} + \dfrac{5}{9} =$ $-\dfrac{4}{7} - \left(-\dfrac{2}{7}\right) =$ $-\dfrac{5}{6} - \left(-\dfrac{1}{6}\right) =$

5 $-\dfrac{3}{8} + 1\dfrac{5}{8} =$ $-\dfrac{3}{14} + 1\dfrac{7}{14} =$ $-\dfrac{2}{3} - \left(-\dfrac{1}{2}\right) =$ $-\dfrac{5}{8} - \left(-\dfrac{1}{5}\right) =$

6 $-\dfrac{9}{12} - \left(-1\dfrac{5}{12}\right) =$ $\dfrac{5}{9} - \left(-3\dfrac{2}{9}\right) =$ $-\dfrac{3}{25} + 2\dfrac{7}{25} =$ $-\dfrac{6}{7} + 2\dfrac{2}{7} =$

Multiplying Fractions

56

⭐ To multiply fractions, find the product of the numerators and the product of the denominators. To simplify, reduce products to lowest terms.

$$\frac{1}{2} \times \frac{2}{3} = \frac{1 \times 2}{2 \times 3} = \frac{2}{6}$$

Reduce to lowest terms by dividing by the LCM of the numerator and denominator.

$$\frac{2 \div 2}{6 \div 2} = \frac{1}{3}$$

Multiply.

1 $\frac{1}{3} \times \frac{1}{2} =$ $\frac{1}{2} \times \frac{1}{5} =$ $\frac{2}{5} \times \frac{1}{6} =$ $\frac{3}{4} \times \frac{1}{3} =$

2 $\frac{5}{6} \times \frac{3}{4} =$ $\frac{1}{8} \times \frac{2}{3} =$ $\frac{1}{2} \times \frac{5}{7} =$ $\frac{2}{3} \times \frac{1}{2} =$

3 $\frac{4}{10} \times \frac{2}{5} =$ $\frac{2}{9} \times \frac{3}{5} =$ $\frac{3}{12} \times \frac{1}{4} =$ $\frac{5}{6} \times \frac{1}{2} =$

4 $\frac{4}{5} \times \frac{2}{3} =$ $\frac{8}{9} \times \frac{5}{6} =$ $\frac{12}{24} \times \frac{2}{3} =$ $\frac{1}{7} \times \frac{8}{10} =$

5 $\frac{1}{4} \times \frac{5}{8} =$ $\frac{3}{17} \times \frac{2}{3} =$ $\frac{5}{9} \times \frac{11}{12} =$ $\frac{3}{16} \times \frac{1}{3} =$

6 $\frac{5}{6} \times \frac{2}{9} =$ $\frac{7}{16} \times \frac{3}{4} =$ $\frac{5}{12} \times \frac{4}{7} =$ $\frac{10}{32} \times \frac{4}{5} =$

Advantage Math Grade 7 © 2005 Creative Teaching Press

Multiplying Fractions

57

⭐ Change factors that are mixed numbers to improper fractions; then multiply. Reduce products to lowest terms.

$$\frac{7}{8} \times 1\frac{1}{3} = \frac{7}{8} \times \frac{4}{3} = \frac{28}{24} = 1\frac{4}{24}$$

$$1\frac{4}{24} = 1\frac{4 \div 4}{24 \div 4} = 1\frac{1}{6}$$

Multiply.

1 $4\frac{3}{4} \times \frac{2}{3} =$ $1\frac{7}{8} \times \frac{2}{3} =$ $3\frac{1}{8} \times \frac{3}{4} =$

2 $\frac{4}{5} \times 1\frac{7}{8} =$ $\frac{2}{3} \times 3\frac{1}{8} =$ $\frac{1}{8} \times 2\frac{5}{6} =$

3 $\frac{1}{4} \times 3\frac{1}{3} =$ $1\frac{1}{2} \times 2\frac{1}{2} =$ $5\frac{1}{4} \times \frac{3}{4} =$

4 $\frac{4}{5} \times 1\frac{1}{2} =$ $\frac{6}{10} \times 2\frac{5}{8} =$ $1\frac{1}{6} \times \frac{9}{13} =$

5 $6\frac{1}{3} \times 4\frac{2}{3} =$ $7\frac{4}{9} \times 3\frac{1}{2} =$ $12\frac{3}{4} \times 2\frac{7}{8} =$

6 $\frac{3}{16} \times 2\frac{7}{8} =$ $3\frac{3}{4} \times \frac{1}{9} =$ $\frac{3}{16} \times 2\frac{2}{5} =$

Dividing Fractions

58

⭐ To divide fractions, invert the divisor to create its reciprocal, and multiply. Reduce quotients to lowest terms.

$$\frac{7}{9} \div \frac{1}{4} = \frac{7}{9} \times \frac{4}{1} = \frac{28}{9}$$

$$\frac{28}{9} = 3\frac{1}{9}$$

Divide.

1. $\dfrac{1}{3} \div \dfrac{1}{4} =$ $\dfrac{1}{8} \div \dfrac{3}{10} =$ $\dfrac{2}{7} \div \dfrac{1}{7} =$

2. $\dfrac{2}{3} \div \dfrac{1}{2} =$ $\dfrac{2}{5} \div \dfrac{1}{4} =$ $\dfrac{3}{8} \div \dfrac{1}{3} =$

3. $\dfrac{2}{3} \div \dfrac{1}{5} =$ $\dfrac{1}{8} \div \dfrac{1}{6} =$ $\dfrac{1}{3} \div \dfrac{1}{4} =$

4. $\dfrac{2}{3} \div \dfrac{2}{5} =$ $\dfrac{12}{21} \div \dfrac{1}{5} =$ $\dfrac{8}{12} \div \dfrac{3}{7} =$

5. $\dfrac{9}{12} \div \dfrac{1}{8} =$ $\dfrac{7}{10} \div \dfrac{3}{5} =$ $\dfrac{5}{12} \div \dfrac{1}{6} =$

6. $\dfrac{3}{10} \div \dfrac{1}{5} =$ $\dfrac{11}{12} \div \dfrac{3}{8} =$ $\dfrac{7}{8} \div \dfrac{5}{6} =$

Dividing Fractions

59

⭐ Change mixed numbers to improper fractions. Reduce quotients to lowest terms.

$$2\frac{7}{8} \div \frac{1}{2} = \frac{23}{8} \times \frac{2}{1} = \frac{46}{8}$$

$$\frac{46}{8} = 5\frac{6}{8} = 5\frac{3}{4}$$

Divide.

1 $1\frac{1}{5} \div \frac{1}{3} =$ $1\frac{2}{3} \div \frac{1}{2} =$ $5\frac{1}{2} \div \frac{2}{3} =$

2 $2\frac{1}{4} \div \frac{7}{8} =$ $5\frac{1}{3} \div \frac{6}{7} =$ $4\frac{2}{7} \div \frac{7}{8} =$

3 $1\frac{3}{10} \div \frac{5}{6} =$ $3\frac{7}{8} \div \frac{1}{3} =$ $6\frac{1}{4} \div \frac{2}{3} =$

4 $2\frac{1}{6} \div \frac{5}{8} =$ $4\frac{1}{2} \div \frac{2}{3} =$ $5\frac{1}{8} \div \frac{7}{8} =$

5 $3\frac{2}{3} \div \frac{1}{5} =$ $2\frac{4}{5} \div \frac{6}{7} =$ $6\frac{2}{3} \div \frac{4}{9} =$

6 $4\frac{2}{7} \div \frac{7}{10} =$ $3\frac{3}{5} \div \frac{5}{7} =$ $2\frac{7}{10} \div \frac{4}{9} =$

Name _____

Multiplying and Dividing Fractions by Integers

60

⭐ To multiply or divide a fraction by an integer, follow the same rules as multiplying two fractions. An integer can be considered to be a fraction with a denominator of 1.

$$\frac{11}{32} \times {}^-4 = \frac{11}{32} \times \frac{{}^-4}{1} = \frac{{}^-44}{32}$$

$$\frac{{}^-44}{32} = -1\frac{12}{32} = -1\frac{3}{8}$$

Multiply or divide.

1 $\dfrac{3}{8} \times {}^-8 =$ 　　　　　 $\dfrac{5}{7} \times {}^-7 =$ 　　　　　 $\dfrac{5}{12} \times {}^-9 =$

2 $\dfrac{7}{10} \times {}^-3 =$ 　　　　　 $\dfrac{5}{12} \times {}^-6 =$ 　　　　　 $\dfrac{4}{5} \times {}^-8 =$

3 $\dfrac{5}{8} \div {}^-3 =$ 　　　　　 $\dfrac{1}{3} \div {}^-8 =$ 　　　　　 $\dfrac{3}{10} \div {}^-7 =$

4 $\dfrac{4}{9} \div {}^-8 =$ 　　　　　 $\dfrac{9}{12} \div {}^-6 =$ 　　　　　 $\dfrac{7}{8} \div {}^-5 =$

5 $\dfrac{19}{36} \div {}^-3 =$ 　　　　　 $\dfrac{5}{6} \times {}^-4 =$ 　　　　　 $\dfrac{7}{12} \div {}^-5 =$

6 $\dfrac{2}{9} \div {}^-6 =$ 　　　　　 $\dfrac{5}{8} \times {}^-2 =$ 　　　　　 $\dfrac{6}{7} \div {}^-12 =$

Advantage Math Grade 7 © 2005 Creative Teaching Press

Multiplication and Division of Fractions Practice

Multiply or divide.

1 $\dfrac{3}{5} \times \dfrac{4}{5} =$ $\dfrac{6}{8} \times \dfrac{3}{8} =$ $\dfrac{2}{5} \times \dfrac{3}{8} =$

2 $\dfrac{2}{3} \div \dfrac{1}{6} =$ $\dfrac{7}{8} \div \dfrac{2}{5} =$ $\dfrac{5}{8} \div \dfrac{1}{6} =$

3 $\dfrac{5}{6} \times \dfrac{3}{4} =$ $\dfrac{1}{2} \times \dfrac{1}{6} =$ $\dfrac{3}{4} \times \dfrac{3}{5} =$

4 $\dfrac{9}{10} \times \dfrac{2}{5} =$ $\dfrac{7}{12} \div \dfrac{1}{4} =$ $\dfrac{4}{5} \div \dfrac{3}{10} =$

5 $3\dfrac{1}{4} \times \dfrac{5}{8} =$ $3\dfrac{1}{6} \times \dfrac{2}{5} =$ $7\dfrac{5}{8} \times \dfrac{2}{5} =$

6 $1\dfrac{1}{10} \times \dfrac{3}{4} =$ $4\dfrac{2}{7} \times \dfrac{6}{8} =$ $5\dfrac{1}{3} \times \dfrac{5}{9} =$

Name _____

62

Multiply or divide.

1 $5\dfrac{1}{3} \div \dfrac{3}{4} =$ $2\dfrac{1}{4} \div \dfrac{2}{3} =$ $5\dfrac{1}{2} \div \dfrac{1}{8} =$

2 $2\dfrac{5}{8} \div \dfrac{7}{9} =$ $1\dfrac{1}{8} \div \dfrac{12}{18} =$ $6\dfrac{2}{3} \div \dfrac{7}{10} =$

3 $2\dfrac{7}{10} \div \dfrac{4}{9} =$ $7\dfrac{4}{9} \times \dfrac{2}{3} =$ $1\dfrac{1}{8} \times \dfrac{9}{13} =$

4 $12\dfrac{3}{4} \times 2\dfrac{7}{8} =$ $3\dfrac{3}{4} \times \dfrac{2}{9} =$ $2\dfrac{2}{5} \div \dfrac{3}{14} =$

5 $\dfrac{3}{10} \times {}^{-}6 =$ $\dfrac{9}{12} \div {}^{-}6 =$ $\dfrac{4}{5} \times {}^{-}1 =$

6 $\dfrac{5}{6} \times {}^{-}2 =$ $\dfrac{6}{7} \div {}^{-}5 =$ $\dfrac{1}{12} \times {}^{-}6 =$

Take a Test Drive

Fill in the bubble beside the correct answer.

1 $\dfrac{11}{100} - \dfrac{3}{100} =$

- ○ $\dfrac{9}{100}$
- ○ $\dfrac{2}{25}$
- ○ $\dfrac{8}{25}$
- ○ $\dfrac{3}{100}$

5 $3\dfrac{1}{3} \times 2\dfrac{1}{2} =$

- ○ $8\dfrac{1}{6}$
- ○ $6\dfrac{1}{3}$
- ○ $5\dfrac{5}{6}$
- ○ $8\dfrac{1}{3}$

2 $\dfrac{1}{7} + \dfrac{3}{5} =$

- ○ $\dfrac{26}{35}$
- ○ $\dfrac{1}{3}$
- ○ $\dfrac{3}{35}$
- ○ $\dfrac{13}{17}$

6 $\dfrac{3}{4} \div \dfrac{1}{2} =$

- ○ $1\dfrac{1}{4}$
- ○ $4\dfrac{1}{6}$
- ○ $1\dfrac{1}{2}$
- ○ $6\dfrac{1}{4}$

3 $-\dfrac{7}{8} + 3\dfrac{3}{8} =$

- ○ $4\dfrac{1}{3}$
- ○ $2\dfrac{1}{3}$
- ○ $4\dfrac{1}{2}$
- ○ $2\dfrac{1}{2}$

7 $6\dfrac{1}{4} \div 2\dfrac{1}{3} =$

- ○ $1\dfrac{3}{4}$
- ○ $2\dfrac{19}{28}$
- ○ $2\dfrac{3}{4}$
- ○ $3\dfrac{19}{28}$

4 $\dfrac{1}{6} \times \dfrac{7}{8} =$

- ○ $\dfrac{4}{21}$
- ○ $\dfrac{1}{8}$
- ○ $\dfrac{7}{48}$
- ○ $\dfrac{3}{24}$

8 $\dfrac{5}{9} \div {}^-5 =$

- ○ $\dfrac{1}{5}$
- ○ $\dfrac{1}{9}$
- ○ $-\dfrac{1}{5}$
- ○ $-\dfrac{1}{9}$

Take a Test Drive

64

Fill in the bubble beside the correct answer.

---

1 $\dfrac{7}{15} \times \dfrac{5}{7} =$

     ○ $\dfrac{1}{4}$ ○ $\dfrac{1}{2}$

     ○ $\dfrac{1}{3}$ ○ $\dfrac{2}{3}$

---

5 $3\dfrac{1}{5} \times 5\dfrac{1}{3} =$

     ○ $15\dfrac{1}{8}$ ○ $17\dfrac{1}{15}$

     ○ $15\dfrac{1}{7}$ ○ $17\dfrac{1}{18}$

---

2 $\dfrac{4}{5} - \dfrac{2}{3} =$

     ○ $\dfrac{2}{15}$ ○ $\dfrac{2}{12}$

     ○ $\dfrac{1}{5}$ ○ $\dfrac{5}{6}$

---

6 $\dfrac{5}{6} \div \dfrac{1}{2} =$

     ○ $\dfrac{6}{10}$ ○ $1\dfrac{3}{5}$

     ○ $1\dfrac{2}{3}$ ○ $\dfrac{5}{12}$

---

3 $25\dfrac{3}{8} - 21\dfrac{5}{8} =$

     ○ $1\dfrac{3}{4}$ ○ $3\dfrac{3}{4}$

     ○ $2\dfrac{3}{4}$ ○ $4\dfrac{3}{4}$

---

7 $2\dfrac{1}{7} \div 1\dfrac{1}{3} =$

     ○ $1\dfrac{1}{2}$ ○ $1\dfrac{8}{14}$

     ○ $1\dfrac{15}{28}$ ○ $1\dfrac{17}{28}$

---

4 $\dfrac{3}{4} \times \dfrac{6}{7} =$

     ○ $\dfrac{18}{27}$ ○ $\dfrac{21}{24}$

     ○ $\dfrac{9}{14}$ ○ $\dfrac{15}{28}$

---

8 $\dfrac{3}{4} \div {}^{-}4 =$

     ○ $\dfrac{1}{3}$ ○ $\dfrac{3}{16}$

     ○ $-\dfrac{1}{3}$ ○ $-\dfrac{3}{16}$

Advantage Math Grade 7 © 2005 Creative Teaching Press

Order of Operations

65

⭐ To evaluate an expression requiring several operations, follow this sequence:

- Do operations in **P**arentheses and other grouping symbols first. If there are grouping symbols within other grouping symbols, do the innermost first.
- Convert **E**xponential numbers to standard numbers.
- **M**ultiply and **D**ivide from left to right.
- **A**dd and **S**ubtract from left to right.

$$12 \times 3 - 2(8 + 4) + 3^2 = ?$$

Think of the acronym **PEMDAS** to remember the order of operations.

$12 \times 3 - 2(8 + 4) + 3^2 = ?$

Parentheses

$12 \times 3 - 2(12) + 3^2 = ?$

Exponents

$12 \times 3 - 2(12) + 9 = ?$

Multiply and Divide

$36 - 24 + 9 = ?$

Add and Subtract

$36 - 24 + 9 = 21$

Evaluate these expressions.

1 $2^2 + 2(3 + 4) \div 9 =$

2 $(16 \div 4) + (14 \div 2) - 3^2 =$

3 $(3^{-1} \times 9)^2 =$

4 $20 \div (2 + 3) + 7 =$

5 $7(8 - 5) + (4 - 1)^2 =$

6 $(6 + 8) \times (6 - 8) =$

7 $(5 \times 4) + (2 \div 2) \times (7 - 4) =$

8 $2(3)^2 + 5(4) - 6 =$

Name _____

Order of Operations

66

Which expression has the greater value? Use < or >.

1 $(5+3) \times 4 + 4^2 \bigcirc 5 + (3 \times 4) + 4^2$

2 $2 \times (1 + 3^2) - 6 + 8 \bigcirc 2 \times 1 + (3^2 - 6) + 8$

3 $(5^2 \times 2) + (6^2 \div 12) \bigcirc 5^2 \times (2 + 6^2) \div 12$

4 $(24 \div 3) \times 2^2 + 9 \bigcirc 24 \div (3 \times 2^2) + 9$

Follow the order of operations and find the missing number.

5 $(3^2 + ?) \div 6 + 4 = 8$

6 $(2^2 + ?) \times (2^2 - 1) = 15$

7 $(2^2 + ?) \div (2^2 - 1) = 15$

8 $14 - (12 \div 2^2) + ? = 2$

9 $(4^2 - 7)^? = \dfrac{1}{9}$

Add parentheses, operation signs, and exponents where they belong.

10 7   6   10   2   = 62

Advantage Math Grade 7 © 2005 Creative Teaching Press

Name _____

Evaluating Mathematical Expressions

67 ⭐ $2x + 3$

A mathematical expression can include numbers, operations signs, and variables. In the expression $2x + 3$, x is a variable because it might have different values. For example, if $x = 3$, the expression $2x + 3$ becomes $2 \times 3 + 3$, which is equal to $6 + 3$ or 9. But when $x = 4$, the expression becomes $2 \times 4 + 3$, which is equal to 11.

To evaluate an expression with a variable, simply substitute the value of the variable into the expression and simplify.

Evaluate for $a = 2$, $b = 6$, $c = 12$.

1 $a + 4 =$ $3b =$ $c^2 =$

2 $b + 4 =$ $3c =$ $a^2 =$

3 $a + c =$ $a + b =$ $b^2 - c =$

Evaluate for each value of d.

	d	$d^2 + 2$	$10 - d$	$1/3\, d$
4	1			
5	2			
6	3			
7	4			

Evaluate for each value of x.

8 $x = \dfrac{1}{2}$ $x = 2$ $x = 4$
   $2x + 3 =$ $2x + 3 =$ $2x + 3 =$

Evaluate for $f = 3$, $g = 2f$, $h = 3f$.

9 $2f =$ $2g =$ $2h =$

10 $f + g =$ $f + h =$ $g + h =$

Name _____

Evaluating Mathematical Expressions

68 Sometimes a mathematical expression has two variables. In the following expression, both x and y are variables.

$$x^2 + y + 3$$

To evaluate an expression with two variables, simply substitute the value of each variable into the expression and simplify.

$x = 3$ and $y = 1$

$x^2 + y + 3 =$

$3^2 + 1 + 3 =$

$9 + 1 + 3 = 13$

Evaluate for $x = 2$ and $y = 3$.

1 $\quad 3x^2 + y = \qquad\qquad x^3 + 2y = \qquad\qquad x^2 - y =$

2 $\quad x - y = \qquad\qquad \dfrac{2(x+y)}{100} = \qquad\qquad 2(y - x) =$

3 $\quad x^2 + y - 3 = \qquad\qquad x^2 + y^2 + 4 = \qquad\qquad x^{-1}y =$

4 $\quad {}^-x - y = \qquad\qquad {}^-x^2 - y^2 = \qquad\qquad y^2 - x^2 =$

Evaluate for $a = 5$ and $b = {}^-3$.

5 $\quad 2a^2 + b = \qquad\qquad b^2 - a = \qquad\qquad a^2 - b =$

6 $\quad a - b = \qquad\qquad b - a = \qquad\qquad 2(a + b) =$

7 $\quad 2a^2 + 3b - 6 = \qquad\qquad 3b^2 - a + 4 = \qquad\qquad a^2 + b^2 + 1 =$

8 $\quad {}^-a - b = \qquad\qquad {}^-a^2 - b^2 = \qquad\qquad b^2 + a =$

Advantage Math Grade 7 © 2005 Creative Teaching Press

Solving Equations

69

An equation is a mathematical sentence containing two expressions separated by an equal sign. The expression on the left of the equal sign has the same value as the expression on the right. One or both of the expressions may contain variables.

To solve an equation, find the value of the variable.

$x - 3 = 10$

To solve for x, add 3 to both sides of the equation.

$x - 3 + 3 = 10 + 3$

$x = 13$

Equations are always balanced, and one side is always equivalent to the other. When solving an equation, remember that whatever you do to one side, you must do to the other.

Solve.

1 $4 + x = 9$ $\qquad\qquad$ $x + 5 = 6$

2 $15 = 20 - a$ $\qquad\qquad$ $18 - a = 5$

3 $b + 3b = 4$ $\qquad\qquad$ $b^2 = 9$

4 $6y = 18$ $\qquad\qquad$ $\dfrac{1}{10}y = 10$

5 $c \div 8 = 32$ $\qquad\qquad$ $c \div 5 = 5$

6 $4x = 36$ $\qquad\qquad$ $2x + 2 = 10$

7 $1.06y = 3.18$ $\qquad\qquad$ $y - \dfrac{1}{2} = 3\dfrac{1}{4}$

8 $a + 2 = {}^-11$ $\qquad\qquad$ $-3 + a = 10$

9 $n \div 2 = {}^-5$ $\qquad\qquad$ ${}^-4 + n = 8$

10 $x - 4 = {}^-2$ $\qquad\qquad$ $x + 1 = 0$

Name _____

Solving Equations

70 ⭐ Solving equations sometimes requires two or three steps.

Solve for x.
$$3x - 4 = 2x$$

First, add 4 to both sides of the equation.
$$3x - 4 + 4 = 2x + 4$$
$$3x = 2x + 4$$

Next, subtract 2x from both sides of the equation.
$$3x - 2x = 2x + 4 - 2x$$
$$x = 4$$

Solve.

1 $2x + 2 = 4$ $3x + 6 = 12$

2 $x^2 + 5 = 30$ $x^2 - 6 = 30$

3 $2y + 6 = 20$ $\dfrac{y}{4} - 9 = 3$

4 $3n - 5 = 16$ $16 = 4 + 3n$

5 $8p + 7 = 55$ $7 + 6p = 31$

6 $\dfrac{a}{5} + 5 = 15$ $\dfrac{a}{3} - 6 = 18$

7 $2b^2 + 2 = 20$ $3b^3 + 3 = 27$

8 $3c + 2 = 20$ $3c^2 + 2 = 50$

9 $38 = 8x + 6$ $5x + 30 = 60$

10 $4y + 4 = 0$ $3y + 1 = {}^-11$

Advantage Math Grade 7 © 2005 Creative Teaching Press

Solving Equations

71

Solve.

1 $14y = 168$ $\qquad\qquad$ $15y = 150$

2 $22a = 242$ $\qquad\qquad$ $21a = 168$

3 $7x - 10 = 25$ $\qquad\qquad$ $3x - 9 = {}^-3$

4 $9y - 4 = 5$ $\qquad\qquad$ $4y - 4 = 4$

5 $x^2 - 9 = 40$ $\qquad\qquad$ $4 + x^2 = 68$

6 $2a^3 + 1 = 17$ $\qquad\qquad$ $3a^2 - 5 = 43$

7 $9b - 1 = 44$ $\qquad\qquad$ $5b - 10 = {}^-5$

8 $5c + 6 = 1$ $\qquad\qquad$ $7 + 6c = 31$

9 $d - {}^-114 = {}^-3207$ $\qquad\qquad$ $d - {}^-1,772 = {}^-435$

10 $1,624 + a = {}^-323$ $\qquad\qquad$ ${}^-2,748 + a = {}^-506$

11 $1,600b = 400$ $\qquad\qquad$ $2,570b = 257$

12 $625x = 25$ $\qquad\qquad$ $25x = 625$

Solving Inequalities

72

An inequality is similar to an equation. Two expressions separated by a symbol that tells how one expression is related to the other. The > sign is used to show that the left side has a greater value than the right side. The < sign shows that the left side has a lesser value than the right side.

Solve for x.

$6x > 30$

Divide both sides of the inequality by 6.

$6x \div 6 > 30 \div 6$

$x > 5$

The solution tells us that for the equation to be true, x could be any number that is greater than 5.

Solve.

1 $\quad 4x - 7 < {}^-3$ $\qquad\qquad 4x - 3 < 9$

2 $\quad 5y + 4 > 19$ $\qquad\qquad 8y - 7 > 49$

3 $\quad 9a - 3 > 69$ $\qquad\qquad 4a - 4 < 31$

4 $\quad 7c - 1 < 41$ $\qquad\qquad 4c + 2 > 6$

5 $\quad 3x - 3 < 3$ $\qquad\qquad 8x + 5 > 53$

6 $\quad 6y + 6 > 24$ $\qquad\qquad 7y - 7 > 7$

7 $\quad 10n + 5 < 45$ $\qquad\qquad 8n - 9 > 39$

8 $\quad 10p + 10 < 90$ $\qquad\qquad 7p + 10 > 17$

9 $\quad 9x - 10 > 44$ $\qquad\qquad 4x - 3 > 1$

 Advantage Math Grade 7 © 2005 Creative Teaching Press

Writing Equations

73

Write an equation, an inequality, or a word problem to match the equation. Then solve. Show your work.

1 Dan wants to build his collection of toy cars so that he has more than his brother, Tom. If Tom has 15 and Dan has 12, how many toy cars, *x*, does Dan need to accomplish his goal?

2 A box of pencils is divided into three batches of *x* pencils each. If 2 pencils are added to each batch, there are 21 pencils. How many pencils were in the box?

3 A town of 1,200 households is divided into four voting districts. Each district has *x* number of households. How many households does each district have?

4 The low temperature for Tuesday was 33 degrees. The high was *x* degrees—just a bit less than 8 degrees higher. What was the high temperature for Tuesday?

5 There are five teams in the league, each with *x* players. After 10 new players are signed up, there are 70 players in all. How many players were on each team before the 10 new players signed up?

6 If a driver doubles her speed to just over 50 mph, what was her speed, *x*, to begin with?

7 Write a word problem to match $2x = 150$.

8 Write a word problem to match $3x - 5 = 25$.

Take a Test Drive

74

Fill in the bubble beside the correct answer.

---

1 Evaluate $9 \times 4 - 3 (6 + 2) + 2^2$.

- ○ 12
- ○ 16
- ○ 20
- ○ 24

5 Choose the answer that best completes the equation: $4^2 - ? + 2 (4 + 3) = 24$.

- ○ 6
- ○ 8
- ○ 10
- ○ 12

---

2 Evaluate $3^2 + 3 (2 + 1) - 9$.

- ○ 27
- ○ 18
- ○ 9
- ○ 3

6 Choose the answer that best completes the equation: $(5^2 - 9)^? = 4$.

- ○ -1
- ○ 0
- ○ ½
- ○ 2

---

3 Evaluate $(4^2 + 2) + (6 \times 4)$.

- ○ 96
- ○ 28
- ○ 34
- ○ 42

7 Evaluate $d^2 + 5$ for $d = 3$.

- ○ 3
- ○ 8
- ○ 11
- ○ 14

---

4 Which best completes the statement?
$(5 + 4)^2 + 2$ ○ $5 + 8 \times 10$

- ○ >
- ○ <
- ○ =
- ○ ≠

8 Evaluate $x^2 - y^2$ for $x = 2$, $y = 5$.

- ○ 21
- ○ -21
- ○ 9
- ○ -9

Advantage Math Grade 7 © 2005 Creative Teaching Press

Take a Test Drive

Fill in the bubble beside the correct answer.

---

1 $4x = 32$

$x =$

○ 8

○ ⅛

○ 28

○ 36

---

2 $2.5n = 6.25$

$n =$

○ 2.5

○ 3.75

○ 6.5

○ 8.75

---

3 $^-4 + a = 0$

$a =$

○ 4

○ ⁻4

○ ¼

○ 0

---

4 $4p + 14 = 30$

$p =$

○ 4

○ 40

○ 42

○ 0.4

---

5 $2n^3 + 4 = 20$

$n =$

○ 1

○ 2

○ 3

○ 4

---

6 Which is true if $5y - 5 > 30$?

○ $y > 7$

○ $y < 7$

○ $y = 7$

○ $y \neq 7$

---

7 If there are 3 dozen pencils in a box, how many boxes are needed for 360 pencils?

○ $36n = 360$ ○ $12n = 360$

○ $36 + n = 360$ ○ $3n \times 12n = 360$

---

8 If a man loses 27 pounds and now weighs 195 pounds, how much did he weigh before?

○ $n - 27 = 195$ ○ $27 - n = 195$

○ $n + 27 = 195$ ○ $27 + n = 195$

Elapsed Time

76

⭐ When adding and subtracting measures of time, keep in mind that 60 seconds equals 1 minute, and 60 minutes equals 1 hour.

Coach Mazzo started the first cheerleading practice session at 8:55. The team worked for an hour and 10 minutes and then took a break. They resumed practicing 30 minutes later and ended at 11:15. How much time did they practice in all?

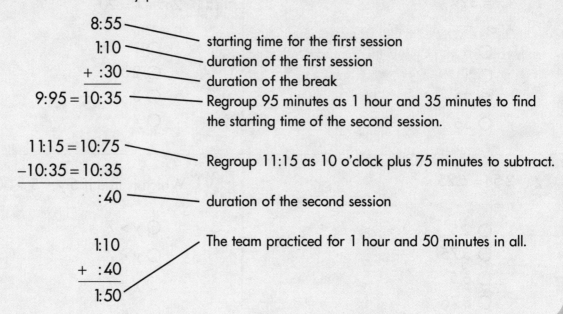

8:55 —— starting time for the first session

1:10 —— duration of the first session

+ :30 —— duration of the break

9:95 = 10:35 —— Regroup 95 minutes as 1 hour and 35 minutes to find the starting time of the second session.

11:15 = 10:75

−10:35 = 10:35 —— Regroup 11:15 as 10 o'clock plus 75 minutes to subtract.

:40 —— duration of the second session

The team practiced for 1 hour and 50 minutes in all.

1:10

+ :40

1:50

Solve.

1 Paz and Belinda left their house at 3:45 and walked for 25 minutes before arriving at the library. What time did they arrive at the library?

2 The girls left the library an hour and 35 minutes later when it closed. Their walk home took another 25 minutes. What time did they arrive home?

3 Frank and Joe want to visit their cousin Sam who lives 4 hours and 15 minutes away. What time should they leave if they want to arrive at 1:30 p.m.?

4 The sun rose at 6:48 a.m. and set at 7:14 p.m. How long was it from sunrise to sunset?

5 Lena crossed the finish line at 12:56. Arlene crossed the finish line 14 minutes later. What time did Arlene finish the race?

Advantage Math Grade 7 © 2005 Creative Teaching Press

Measuring Temperature

77

In the U.S. Customary System, the Fahrenheit scale is used to measure temperature. On a Fahrenheit thermometer, 32° is the temperature at which water at sea level freezes, and 212° is the temperature at which it boils.

In the metric system, the Celsius scale is used to measure temperature. On a Celsius thermometer, water freezes at 0° and boils at 100°.

Read the thermometers and write the temperature.

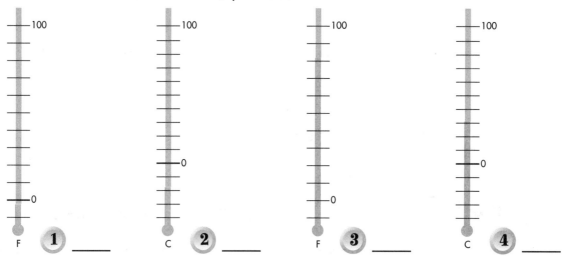

1 _____   2 _____   3 _____   4 _____

Solve. Show your work.

5 The temperature in a cooler dropped to freezing and then continued to drop another 7 degrees. What temperature was it in the cooler on a Fahrenheit thermometer?

6 The weather forecast called for the day's high to reach 11° Celsius and the low to reach 3° below freezing. How great is the day's temperature range?

7 The temperature in a freezer that was set at 0° Fahrenheit started rising because the electricity went off. It rose until the ice cubes thawed. How much did the temperature rise?

Measuring Weight or Mass

78

⭐ In the U.S. Customary System, weight is a measure of the pull of gravity on the matter in an object on Earth. Weight is measured using ounces (oz.), pounds (lb.), and tons (T.).

16 oz. = 1 lb.

2,000 lb. = 1 T.

In the metric system, mass is the amount of matter in an object. Mass and weight are similar but not exactly the same thing. Gravity affects weight but not mass. Mass is measured using milligrams (mg), grams (g), kilograms (kg), and metric tons (t).

1 mg = 0.001 g

1 kg = 1,000 g

1,000 kg = 1 t

Solve.

1 72 oz. = ____ lb.   4 lb. = ____ oz.   100 oz. = ____ lb.

2 $2\frac{1}{2}$ lb. = ____ oz.   6 kg = ____ g   400 mg = ____ g

3 0.32 kg = ____ g   0.28 g = ____ mg   10,000 kg = ____ t

Complete each sentence using <, >, or =.

4 1 lb. ◯ 14 oz.   100 mg ◯ 1 g   1 t ◯ 10.000 g

5 2 T. ◯ 400 lb.   1,500 lb. ◯ $\frac{1}{2}$ T. $\frac{1}{4}$ T. ◯ 500 lb.

6 5,200 mg ◯ 52 g   2,000 kg ◯ 2 t   180 g ◯ 0.2 kg

     Advantage Math Grade 7 © 2005 Creative Teaching Press

Converting Between U.S. Customary and Metric

79

⭐ Use these formulas to convert between U.S. Customary System and metric system measurements.

Temperature

$$F = \frac{9}{5} C + 32$$

$$C = \frac{5}{9} (F - 32)$$

Weight or Mass

1 oz. = 28.35 g	1g = 1/28.35 oz.
1 lb. = 453.59 g	1 kg = 2.204 lb.
100 lb. = 45.349 kg	100 kg = 220.46 lb.

Solve.

1 5 oz. = ____ g   68°F = ____ °C   20 oz. = ____ g

2 340.2 g = ____ oz.   100°C = ____ °F   25 lb. = ____ kg

3 20°C = ____ °F   1,000 lb. = ____ kg   50 kg = ____ lb.

4 74 lb. = ____ kg   5 g = ____ oz.   75°F = ____ °C

5 2,000 lb. = ____ kg   200 kg = ____ lb.   0°F = ____ °C

6 10 g = ____ oz.   10,000 kg = ____ lb.   1,500 lb. = ____ kg

Name _____

Measuring Length

80

⭐ In the U.S. Customary System, length is measured using inches (in.), feet (ft.), yards (yd.), and miles (mi.).

12 in. = 1 ft.
3 ft. = 1 yd.
5,280 ft. = 1 mi.

In the metric system, length is measured using millimeters (mm), centimeters (cm), meters (m), and kilometers (km).

10 mm = 1 cm
10 cm = 1 m
1,000 m = 1 km

Solve.

1 3,000 m = _____ km   7 ft. 6 in. = _____ yd.   4.2 km = _____ m

2 90 in. = _____ yd.   1 km = _____ cm $4\frac{1}{4}$ ft. = _____ in.

3 0.7 m = _____ mm   2,640 ft. = _____ mi.   1 yd. 2ft. = _____ in.

Complete each sentence using <, >, or =.

4 36 m ◯ 3.6 m   2 yd. ◯ 70 in.   310 cm ◯ 31 m

5 0.8 m ◯ 800 mm   4 mi. ◯ 22,000 ft.   48 in. ◯ 4.8 ft.

6 $3\frac{1}{2}$ yd. ◯ 10 ft.   7.07 m ◯ 707 cm   44 cm ◯ 440 mm

Advantage Math Grade 7 © 2005 Creative Teaching Press

Name _____

Measuring Capacity

81

In the U.S. Customary System, capacity is measured using cups (c.), pints (pt.), quarts (qt.), and gallons (gal.).

1 pt. = 2 c.
1 qt. = 2 pt.
1 gal. = 4 qt.

In the metric system, capacity is measured using milliliters (mL) and liters (L).

1000 mL = 1 L

Solve.

1 $4\frac{1}{2}$ qt. = ____ pt.   40 qt. = ____ pt.   40 pt. = ____ gal.

2 2,500 mL = ____ L   64 qt. = ____ gal.   4.34 L = ____ mL

3 12 pt. = ____ gal.   1.5 L = ____ mL $\frac{3}{4}$ c. = ____ pt.

Complete each sentence using <, >, or =.

4 4 c. ◯ 3 pt.   16 pt. ◯ 1.5 gal.   40 gal. ◯ 150 qt.

5 500 mL ◯ 0.25 L   10 mL ◯ 0.01 L   20 c. ◯ 4 qt.

6 25 c. ◯ 1 gal.   475 mL ◯ 4.75 L   8 pt. ◯ 1 gal.

Converting Between U.S. Customary and Metric

82

⭐ Formulas are used to convert between U.S. Customary System and metric measurements.

Length
1 in. = 2.54 cm
1 ft. = 30.48 cm
1 yd. = 91.44 cm
1 mi. = 1,609.34 m
1 mi. = 1.609 km

Capacity
1 pt. = 473 mL
1 qt. = 946 mL
1 gal. = 3.785 L

Solve.

1 3.4 pt. = _____ mL   7.57 L = _____ qt.   500 mL = _____ pt.

2 5 ft. = _____ m   10 L = _____ pt.   1,500 mL = _____ qt.

3 3.2 mi. = _____ m   33 in. _____ cm   3,219 m = _____ mi.

4 1 yd. = _____ m   50 cm = _____ in.   5.6 km = _____ mi.

5 40 qt. = _____ L   6 L = _____ gal.   1,000 mL = _____ pt.

6 4 qt. = _____ mL   12 cm = _____ in.   10 L = _____ gal.

Naming Angles

83

⭐ An angle is formed by two rays with a common end point.

This angle is called ∠A or ∠BAC.

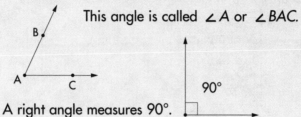

A right angle measures 90°.

An acute angle measures between 0° and 90°.

An obtuse angle measures between 90° and 180°.

A straight angle measures 180°.

A reflex angle measures between 180° and 360°.

Write two names for each angle, and tell what type of angle it is.

1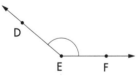

2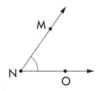

Draw and label a right angle, an acute angle, and an obtuse angle.

3

Complementary and Supplementary Angles

84

⭐ Complementary angles are any two angles whose sum is 90°.

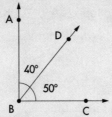

Angle ABD is the complement of angle DBC.

Supplementary angles are any two angles whose sum is 180°.

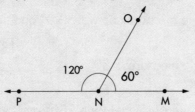

Angle MNO is the supplement of angle ONP.

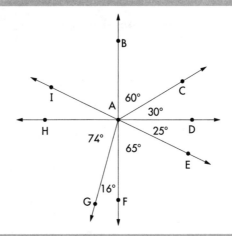

Give the name and measure of each.

	Complement		Supplement	
	name	measure	name	measure
∠ DAF				
∠ EAF				
∠ BAC				
∠ HAG				
∠ EAG				

Perimeter

85

⭐ The perimeter of a figure is the length around its sides.

A square is a four-sided figure in which all sides are of equal length. Therefore, $P = 4s$.

A rectangle is a four-sided figure in which each pair of sides is the same length. Therefore, $P = 2l + 2w$.

An equilateral triangle is a three-sided figure in which all sides are the same length. Therefore, $P = 3s$.

S	L W	S
P=4S	P=2L+2W	P=3S

Find the perimeter of each figure.

1   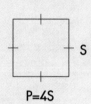 3 cm   6 m 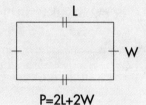 2 m    1 in.

_____   _____   _____

2    4 in.   4 in. 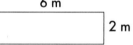 3 in.   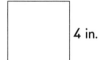 4 in.

_____   _____   _____

3   6 ft.   5 cm  4 cm   2 cm

_____   _____   _____

4    10 in.   7 yd. 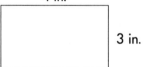 1 yd.   2 ft.

_____   _____   _____

Name _____

Circumference

86 ⭐ Circumference is the distance around a circle. The diameter is a chord passing through the center. Pi (π) is the ratio of the circumference to the diameter. This ratio has the same value for all circles.

$$\pi = \frac{c}{d}$$

$$\pi = \frac{22}{7} \text{ or about } 3.14$$
$$c = \pi d$$
$$c = 3.14d$$

Find the circumference of each figure.

1
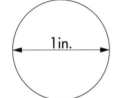
1 in.

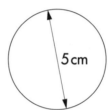
10 in.

5 cm

2
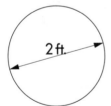
2 ft.

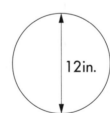
6 cm

12 in.

3

14 mm

3.75 in.

4.2 yd.

Advantage Math Grade 7 © 2005 Creative Teaching Press

Triangles

87 ⭐ The sum of the angles in any triangle is 180°.

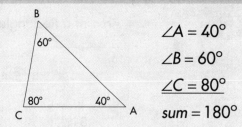

$\angle A = 40°$
$\angle B = 60°$
$\underline{\angle C = 80°}$
$sum = 180°$

A right triangle contains a right angle. The sides of a right triangle have special names. The shorter sides are called legs, and the longer side is called the hypotenuse.

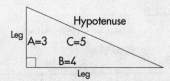

If you know the lengths of the legs, you can find the length of the hypotenuse using the Pythagorean thoerem, which states:

$$a^2 + b^2 = c^2$$

Find the measure of the missing angle.

1

20°

110°

65°

55°

35°   35°

_____   _____   _____

2

70°   70°

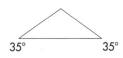

130°

15°

50°   50°

_____   _____   _____

Find the length of the missing side.

3

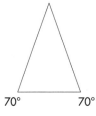

10 in.

24 in.

20 cm   12 cm

8 ft.   15 ft.

Name _____

88

⭐ Area of a Rectangle

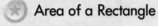

Area = length × width
A = lw
A = 8 in. × 5 in. = 40 sq. in.

5 in.

8 in.

Area of a Square

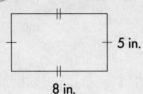

6 cm

Area = side × side
A = s²
A = 36 sq. cm

Area of a Parallelogram

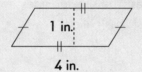

1 in.

4 in.

Area = base × height
A = bh
A = 4 in. × 1 in. = 4 sq. in.

Find the area of each figure.

1

7 cm

8 in.

4 yd.

_____   _____   _____

2

12 in.

 1 in.

4 ft.

 6 ft.

3 ft.   5 ft.

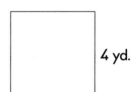 2 ft.

_____   _____   _____

3

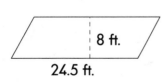

8 ft.

24.5 ft.

4 cm

9 cm

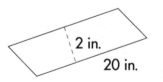

2 in.

20 in.

_____   _____   _____

Advantage Math Grade 7 © 2005 Creative Teaching Press

Name _____

Area of a Circle

89

Area of a Circle
A = π × radius × radius
$A = \pi r^2$
$A = 3.14 \times (5 \text{ in.})^2 = 78.5 \text{ sq. in.}$

5 in.

10 in.

Find the area of each figure.

1

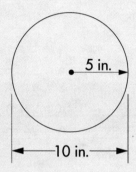

7 cm

4 in.

10 in.

_____ _____ _____

2

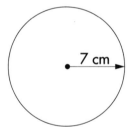

100 mm

15 cm

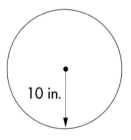

12 in.

_____ _____ _____

3

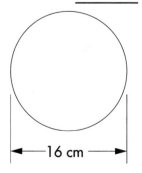

16 cm

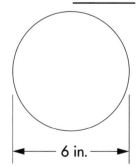

6 in.

2 ft.

_____ _____ _____

Name _____

90

⭐ **Area of a Triangle**

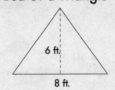

A = ½ × base × height
A = ½ bh
A = ½ × 6 ft. × 8 ft. = 24 sq. ft.

Area of a Trapezoid

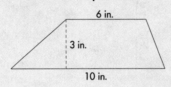

A = ½ × height × (base₁ + base₂)
A = ½ h (b₁+b₂)
A = ½ × 3 in. (6 in. + 10 in.)
A = 24 sq. in.

Find the area of each figure.

1

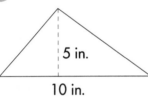

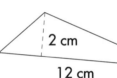

_____   _____   _____

2

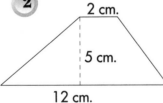

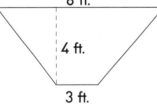

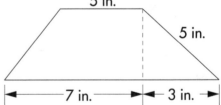

_____   _____   _____

Areas of Irregular Figures

91

Use what you know about the areas of parallelograms, triangles, and trapezoids to find the areas of these figures.

1

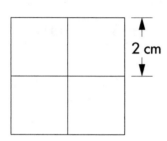
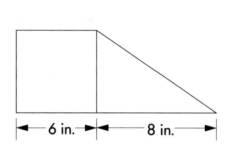
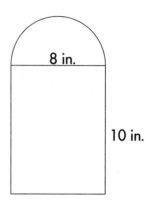

2

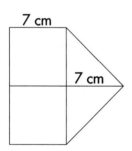
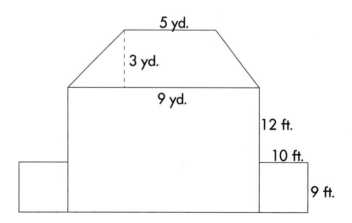

Name _____

Surface Area and Volume of Rectangular Prisms

92

⭐ Surface Area
To find the surface area add the areas of all the faces: front, back, side1, side 2, top, and bottom. Or, use the formula:

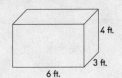

SA = 2lw + 2lh + 2wh
SA = 2(6 in. × 3 in.) + 2(6 in. × 4 in.) +
     2(3 in. × 4 in.)=108 ft²

Volume

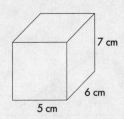

Volume = length × width × height
V = lwh
V = 5 cm × 6 cm × 7 cm = 210 cm³

Find the surface area and volume of each figure.

1

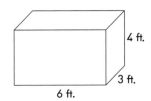

2

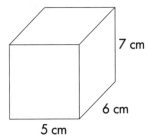

3

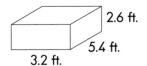

Advantage Math Grade 7 © 2005 Creative Teaching Press

Surface Area and Volume of Triangular Prisms

93

⭐ Surface Area
To find the surface area, add
the areas of all the faces: side 1,
side 2, side 3, top, and bottom.

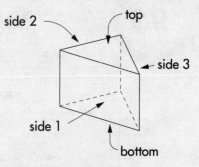

Volume

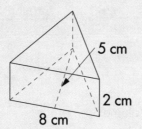

Volume = base area x height

$$V = Bh$$

$$V = (\frac{1}{2}bh_1)h_2$$

$$V = \frac{1}{2} \times 8 \text{ cm} \times 5 \text{ cm} \times 2 \text{ cm} = 40 \text{ cm}^3$$

Find the surface area and volume of each figure.

1

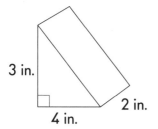

2

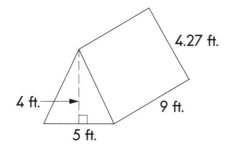

Name _____

Surface Area and Volume of Cylinders

94

⭐ Surface Area
To find the surface area, add the areas of all the faces: top, bottom, and the curved face. Or use the formula:

$$SA = 2\pi r^2 + 2\pi rh$$
$$SA = (2 \times 3.14 \times 1.25 \text{ in.} \times 1.25 \text{ in.}) + (2 \times 3.14 \times 1.25 \text{ in.} \times 4 \text{ in.})$$
$$SA = 41.2125 \text{ in.}^2$$

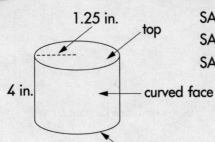

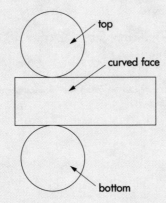

Volume
Volume = base area x height

$$V = Bh$$
$$V = \pi r^2 h$$
$$V = 3.14 \times 1.25 \text{ in.} \times 1.25 \text{ in.} \times 4 \text{ in.}$$
$$V = 19.625 \text{ in.}^3$$

Find the surface area and volume of each figure.

1

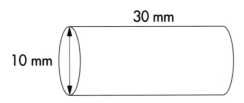

30 mm

10 mm

2

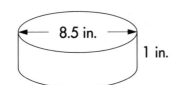

8.5 in.

1 in.

Advantage Math Grade 7 © 2005 Creative Teaching Press

Take a Test Drive

95

Fill in the bubble beside the correct answer.

1 Tia and Gwen swam for 40 minutes, took a 15 minute break, and then went back in the water for another 35 minutes. If they started at 12:10, what time did they finish?

- ○ 1:00
- ○ 1:05
- ○ 1:25
- ○ 1:40

2 1 oz. = _____

- ○ about 56 g
- ○ about 454 g
- ○ about 28 g
- ○ about 2.2 g

3 2 mi. = _____

- ○ about 3,200 m
- ○ about 3,200 km
- ○ about 320 m
- ○ about 32 km

4 2.2 pt. = _____

- ○ about 1 L
- ○ about 100 mL
- ○ about 220 mL
- ○ about 2 L

5 Which describes this angle?

- ○ acute
- ○ right
- ○ obtuse
- ○ reflex

6 Which could be the measure of this angle?

- ○ 95°
- ○ 180°
- ○ 190°
- ○ 280°

7 What is the size of the complement of a 65° angle?

- ○ 35°
- ○ 25°
- ○ 115°
- ○ 295°

8 Which is supplementary to a right angle?

○   ○

○   ○

Name _____

1 What is the perimeter of a rectangle whose length is 5 feet and whose width is 3.5 feet?

○ 8.5 ft.
○ 16.5 ft.
○ 17 ft.
○ 17.5 ft.

2 Given a square with a side of 1m and an equilateral triangle with a side of 4 m, the triangle has a perimeter that is how much longer than the square?

○ 2 times   ○ 4 times
○ 3 times   ○ 5 times

3 What is the perimeter of this figure?

○ 9 in.
○ 12 in.
○ 15 in.
○ 18 in.

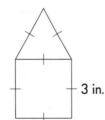

3 in.

4 What is the circumference of a circle whose radius is 3.14 inches?

○ 9.86 in.
○ 19.72 in.
○ 6.28 in.
○ 30.96 in.

5 Which is true of the third angle of a triangle whose other two angles measure 30° and 40°?

○ It is acute.
○ It is a right angle.
○ It is obtuse.
○ It is a reflex angle.

6 What is the length of a leg of a right triangle whose other leg is 15 cm and whose hypotenuse is 25 cm?

○ 850 cm
○ 29 cm
○ 20 cm
○ 15 cm

7 What is the area of this figure?

○ 12.56 in.²
○ 25.12 in.²
○ 50.24 in.²
○ 200.96 in.²

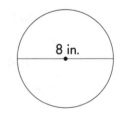

8 in.

8 What is the area of this figure?

○ 34 cm²
○ 38 cm²
○ 40 cm²
○ 46 cm²

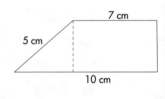

7 cm

5 cm

10 cm

Name _____

Practice Test

97

Fill in the bubble beside the correct answer.

1 Which is the standard form for 54,000,000 + 495,000 + 30 + 1?
- ○ 54,095,301
- ○ 54,495,310
- ○ 54,495,301
- ○ 54,495,031

5 Which is the standard form for 8.3×10^{-5}?
- ○ 0.000083
- ○ 0.00083
- ○ 830,000
- ○ 83,000

2 Which is equivalent to 876,000?
- ○ 8.76×10^6
- ○ 8.76×10^5
- ○ 8.76×10^7
- ○ 876×10^7

6 Which is equivalent to $10^{-3} \times 10^5$?
- ○ 10^{-2}
- ○ 10^2
- ○ 10^{-8}
- ○ 10^8

3 Which is equivalent to $-2(-4^2)$?
- ○ $^-64$
- ○ $^+64$
- ○ $^-32$
- ○ $^+32$

7 Which is equivalent to $\frac{5}{8}$?
- ○ 0.375
- ○ 0.525
- ○ 0.625
- ○ 0.875

4 Which is the square root of 49?
- ○ 5
- ○ 6
- ○ 7
- ○ 8

8 Which is the best estimate of the sum 132 + 488?
- ○ 400
- ○ 500
- ○ 600
- ○ 700

Practice Test

98

Fill in the bubble beside the correct answer.

---

1 $2,009 + 1,296 =$
- ○ 3,205
- ○ 3,295
- ○ 3,305
- ○ 3.405

5 Which property of addition is shown?
$(0.5 + 3.2) + 54 = 0.5 + (3.2 + 54)$
- ○ associative
- ○ commutative
- ○ distributive
- ○ identity

---

2 $0.625 + 0.5 + 1.25 =$
- ○ 0.750
- ○ 0.795
- ○ 1.655
- ○ 2.375

6 $45\overline{)2947.5}$
- ○ 59
- ○ 59.5
- ○ 65
- ○ 65.5

---

3 $6,006 - 997 =$
- ○ 5,009
- ○ 5,019
- ○ 6,009
- ○ 7,003

7 $^{+}105 \div {}^{-}5 =$
- ○ $^{-}21$
- ○ $^{+}21$
- ○ $^{-}25$
- ○ $^{+}25$

---

4 $x + 83 = 75$
$x =$
- ○ $^{+}8$
- ○ $^{-}8$
- ○ $^{+}12$
- ○ $^{-}12$

8 What is the least common multiple of 6 and 9?
- ○ 15
- ○ 18
- ○ 27
- ○ 36

Advantage Math Grade 7 © 2005 Creative Teaching Press

Name _____

99

Fill in the bubble beside the correct answer.

1 $\frac{2}{3} - \frac{1}{4} =$

○ $\frac{5}{12}$ ○ $\frac{3}{8}$

○ $\frac{11}{12}$ ○ $\frac{5}{8}$

2 $\frac{1}{4} \times 2\frac{1}{3} =$

○ $\frac{5}{12}$ ○ $\frac{3}{4}$

○ $\frac{7}{12}$ ○ $2\frac{1}{12}$

3 $\frac{15}{45} = \frac{1}{x}$

$x =$ ○ $\frac{1}{3}$ ○ $\frac{1}{4}$

○ 3 ○ 4

4 $5\frac{1}{2} \div \frac{1}{2} =$

○ $1\frac{1}{11}$ ○ 6

○ $2\frac{3}{4}$ ○ 11

5 $\frac{2}{3} \div {^-3} =$

○ $\frac{2}{9}$ ○ 2

○ $\frac{^-2}{9}$ ○ ⁻2

6 $4^{-1} \times 4^2 =$

○ $\frac{1}{4}$ ○ 16

○ 4 ○ 32

7 Pass Mountain is 3,312 feet tall. The Wind Caves are less than 80 feet from the peak. To what height would you climb to visit the Wind Caves?

○ 3,232 ft. + 80 ft. = x; x = 3,392 ft.

○ 3,312 ft. − 80 ft. = x; x = 3,232 ft.

○ 3,312 ft. − x > 8 ft.; x > 3,232 ft.

○ 3,312 ft. − x < 80 ft.; x < 3,232 ft.

8 Jesse and Kelsey started watching a movie at 6:45. If the movie is 1 hour and 35 minutes long, what time will it end?

○ 7:15

○ 8:20

○ 7:35

○ 8:00

Name _____

100

Fill in the bubble beside the correct answer.

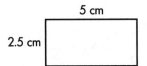

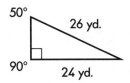

1 What is the perimeter of the rectangle above?
- ○ 7.5 cm
- ○ 12.5 cm
- ○ 15 cm
- ○ 25 cm

4 What is the measure of the missing angle above?
- ○ 30°
- ○ 40°
- ○ 45°
- ○ 90°

2 What is the area of the rectangle above?
- ○ 12.5 cm
- ○ 12.5 cm²
- ○ 15 cm
- ○ 25 cm²

5 What is the length of the missing side above?
- ○ 10 yd.
- ○ 20 yd.
- ○ 40 yd.
- ○ not enough information

3 Which equation shows how to find the area of a square that has a side measuring 3 feet?
- ○ 3 ft. + 3 ft. = 6 ft.
- ○ 3 ft. × 4 ft. = 12 ft.
- ○ 3 ft. × 3 ft. = 9 ft.²
- ○ not enough information

6 Which equation shows how to find the area of a circle with a 6 m diameter?
- ○ 6 m + 6 m = 12 m²
- ○ 3m × 3.14 = 9.42 m²
- ○ 6 m × 3.14 = 18.84 m²
- ○ 9 m² × 3.14 = 28.26 m²

Name _____

Math Grade 7 Tracking Sheet

Activity	Possible	My Score
Unit 1		
1	16	
2	17	
3	18	
4	24	
5	24	
6	20	
7	20	
8	20	
9	16	
10	20	
11	24	
12	25	
13	16	
14	20	
15	36	
Test Scores		
16	8	
17	8	
Unit 2		
18	10	
19	20	
20	14	
21	20	
22	14	
23	20	
24	20	
25	19	
26	20	
27	14	

Activity	Possible	My Score
28	14	
29	20	
30	12	
31	20	
32	15	
33	8	
34	16	
35	18	
36	10	
37	24	
38	16	
39	16	
40	16	
41	24	
42	16	
43	13	
44	13	
45	14	
46	15	
47	20	
48	24	
Test Scores		
49	8	
50	8	
Unit 3		
51	16	
52	16	
53	16	
54	24	
55	24	

Activity	Possible	My Score
56	24	
57	18	
58	18	
59	18	
60	18	
61	18	
62	18	
Test Scores		
63	8	
64	8	
Unit 4		
65	8	
66	10	
67	30	
68	24	
69	20	
70	20	
71	24	
72	18	
73	8	
Test Scores		
74	8	
75	8	
Unit 5		
76	5	
77	7	
78	18	
79	18	
80	18	
81	18	

Activity	Possible	My Score
82	18	
83	7	
84	20	
85	12	
86	9	
87	9	
88	9	
89	9	
90	6	
91	5	
92	6	
93	4	
94	4	
Test Scores		
95	8	
96	8	
97	8	
98	8	
99	8	
100	6	

Math Grade 7 Answer Key

Activity 1
1. 105,932
2. 642,000,300,000
3. 158,006,005
4. 437,003,000,000
5. 53,205,000
6. 797,619,000,132
7. 11,006,007
8. 84,000,037,800
9. 4 thousands   4 ten millions
10. 4 hundreds   4 hundred billions
11. 4 ten millions   4 millions
12. 4 billions   4 hundreds

Activity 2
1. <   >
2. <   >
3. >   <
4. >   <
5. >   >
6. 879 897 978 987
7. 58,683   68,638   83,586
8. 107,432,014   407,210,740   423,413,201
9. 8,400,325,831   8,400,327,937   8,437,293,310
10. 73,302,670   73,603,270   73,627,003
11. Thursday
12. Wednesday

Activity 3
1. 5   3   7
2. 3   5   120
3. 0   18   12
4. -3   -7   -12
5. -25   512   -1296
6. 4   -64   -27

Activity 4
1. 7×10^2 6.5×10^2 1.3×10^3
2. 1.37×10^5 9.87×10^4 1.4×10^6
3. 8×10^6 1×10^7 3.76×10^7
4. 9.875×10^6 6.8×10^{10} 2.593×10^{10}
5. 150   80,000   63
6. 1,600   178,000   104
7. 76,900,000   2,200,000,000   1,400,000
8. 47,800   276,000   187,000,000

Activity 5
1. 1.7×10^{-2} 4.62×10^{-3} 8.74×10^{-3}
2. 3×10^{-5} 7.8×10^{-5} 8.376×10^{-1}
3. 4.2×10^{-6} 1.057×10^{-3} 3.608×10^{-4}
4. 8×10^{-8} 6.403×10^{-4} 4.28×10^{-8}
5. 0.77   0.0043   0.0592
6. 0.000000106   0.0000563   0.00947
7. 0.000007463   0.000000000087   0.000367
8. 0.0000000864   0.000000005397   0.000000000284

Activity 6
1. 81   49   100   9
2. 5   6   10   8
3. 144   64   225   400
4. 7   20   15   14
5. 25   1,024   50   15,625

Activity 7
1. 8   64   5   7,776
2. 512   12   1,024   1
3. 4   4   512   100,000
4. 6   27   0.01   4
5. 1   64   0   7

Activity 8
1. 3^4 5^4 10^{16} 2^{10}
2. 8^5 7^{15} 4^6 3^6
3. 10^{10} 26^{15} 32^6 18^8
4. 2^5 4^3 3^{16} 1^{32}
5. 10^3 6^4 15^8 32^{10}

Activity 9
1. 3^{-8} 6^{-2} 10^{-10} 12^{-2}
2. 7^{-1} 48^{-3} 7^4 6^{-8}
3. 16^{-10} 9^{-10} 1 36^{-3}
4. 5   8   10   3

Activity 10
1. 4/6   16/20   12/16   10/16
2. 1/8   33/48   27/30   18/45
3. 3/4   2/3   2/3   1/2
4. 1/9   1/6   1/5   1/2
5. 1/3   1/8   1/4   3/8

Activity 11
1. 0.375   0.8   0.583
2. 0.167   0.429   0.933
3. 0.9   0.471   0.857
4. 0.625   0.231   0.281
5. 1/4   4/25   2/25
6. 1/100   17/20   1/3
7. 16/25   9/20   1/50
8. 1/20   27/100   49/50

Activity 12
1. 23%   33%   62%   75%
2. 88%   110%   81%   135%
3. 25%   56%   23%   1,072%
4. 3%   58%   0.6%   33%
5. 0.37% 17%   100.1%   1,001%

20%	0.2	1/5
80%	0.8	4/5
15%	0.15	3/20
35%	0.35	7/20
42%	0.42	21/50

Activity 13
1. 900   300   1,100   1,100
2. 600   300   400   100
3. 1,600   1,100   800   1,100
4. 400   300   300   100

Advantage Math Grade 7 © 2005 Creative Teaching Press

Activity 14
1. 673,008,000,000   10,005,002
2. 179,002,005   137,987,000
3. >   >
4. >   >
5. 4   2   8
6. -4   -36   -125
7. 268,000   190,000,000
   67,300
8. 65,000   0.00947
   0.000003647

Activity 15
1. 5   64   14   1024
2. 3   32   0   10,000,000
3. 2^6 6^3 8^{17} 2^{32}
4. 7 4^{-10} 7   10
5. ⅞   ⅓   ½ $^{13}/_{16}$
6. 0.833   0.333   0.75   0.286
7. $^{17}/_{20}$ $^{49}/_{50}$ $^9/_{20}$ $^{16}/_{25}$
8. 0.4%   58.3%   160%   16.3%
9. 25%   27%   4.6%   31%

Activity 16
1. 163,007,504,706
2. 1,743,871
3. -16
4. 3.57×10^7
5. 0.000063
6. 5
7. 8^7
8. 6^4

Activity 17
1. 9^{-10}
2. 72
3. $^{27}/_{72}$
4. 0.80
5. $^{17}/_{20}$
6. 600.3%
7. 87.5%
8. 600

Activity 18
1. commutative
2. associative
3. associative
4. commutative
5. identity
6. commutative
7. identity
8. commutative
9. distributive
10. associative

Activity 19
1. 768   1,123   4,581   9,010
2. 8,769   554   14,800   626
3. 2,606   3,657   10,063   2,399
4. 7,790   4,834   3,177   1,074
5. 10,725   4,467   5,125   3,081

Activity 20
1. 41,703   100,304
   220,832   1,626,921
2. 7,372,119   2,625,492
   4,646,683   9,696,813
3. 12,203,836   3,151,057
   13,902,596   5,961,021
4. 80,503 tickets
5. 5,030,234 cartons

Activity 21
1. 88.36   145.75
   638.801   9.291
2. 266.907   155.389
   1.071   0.00235
3. 1,712.012   1.4948
   4,529.61   8.096
4. 140.1007   26,729.4
   915,029.92   1.431
5. 0.08378   200.5988
   945.757   1,732.873

Activity 22
1. 290.3   0.01822
   791.842   906.479
2. 250.5191   1,881.903
   12.879   2.223
3. 1,819,504.38   0.16721
   1,904.2606   1.0481
4. 38.178 ounces
5. $133.72

Activity 23
1. +20   +48   +13   +28
2. +3   +49   0   +5
3. +180   +200   +120   0
4. +301   +800   +615   +955
5. +7   +39   +1.13   0

Activity 24
1. 1,510   2,124
   795   8,011
2. 12,244   17,943
   89,666   6,480,397
3. 9,072,083   15,427,935
   7579.33   0.08674
4. 190.66   0.019755
   10.847   4,125.13
5. 842.747   1.4063
   10.539   0.00566

Activity 25
1. +24   +57   +39   +30
2. +37   +11.3   +865   0
3. +¾   +⅓   +1   +⅛
4. +0.137   +0.012   +2   +0.38
5. 236,048 employees
6. 0.37 inches
7. $8.77

Activity 26
1. 567   133   533   49
2. 1,030   5,723   2,051   3,780
3. 1,317   1,239   6,405   5,035
4. 4,646   3,889   2,000   2,484
5. 5,082   39   291   876

Activity 27
1. 12,039   96,564
   59,558   30,223
2. 4,555   1,077
   173,849   749,524
3. 8,134,217   268,425
   2,749,025   2,189,089
4. 15,744 hot dogs
5. 7,453,642 people

Activity 28
1. 8.92   63.65   7.65   227.05

2. 85.468   34.497   74.393
   9.246
3. 88.09   752.449   1,231.644
   2,591.499
4. $72.37
5. 4,126.437 pounds

Activity 29
1. -3   3   -4   18
2. -40   22   12   0
3. -37   37   -37   94
4. 2.9   -3.5   31   8.4
5. ¼   -½   ⁵⁄₁₂   -⅔

Activity 30
1. 4   32   28
2. 45   59   127
3. 24   199   138
4. 1,756 1,903 1,155

Activity 31
1. 1,457   9,145   664   1,507
2. 67,017   11,626
   4,646,158   3,970116
3. 16.73   79   72.839   11.917
4. 1,373.78   657.907
   809.049   1,374.945
5. -57   52   0   6

Activity 32
1. 1.8   -1.4   28.1   25.94
2. x=51   x=11   x=26   x=1,040
3. x=100   x=180   x= ¾
   x= ⅝
4. 579,350 beans
5. 363 cards
6. 18.7 miles

Activity 33
1. 1, 2, 3, 6, 9, 18 1, 3, 5, 9, 15, 45
2. 1, 2, 3, 4, 6, 9, 12, 18, 36
   1, 7, 49
3. 8, 16, 24, 32, 40
   9, 18, 27, 36, 45
4. 11, 22, 33, 44, 55
   20, 40, 60, 80, 100

Activity 34
1. 1, 2, 3, 4, 6, 8, 12, 24 1, 2, 4, 8,
   16, 32   circle 8 in each
2. 1, 3, 5, 15   1, 3, 9, 27
   circle 3 in each
3. 1, 2, 4, 8, 16, 32, 64 1, 2, 3, 5, 6,
   10, 15, 30 circle 2 in each
4. 1, 2, 5, 10, 25, 50
   1, 2, 3, 4, 6, 8, 12, 16, 24, 48
   circle 2 in each
5. 8   7
6. 8   25
7. 15   27
8. 7   9

Activity 35
1. prime 1, 7
   composite 1, 3, 5, 15
   composite 1, 2, 4, 8
2. composite 1, 2, 4, 8, 16, 32
   prime 1, 37
   composite 1, 2, 3, 6, 9, 18
3. composite 1, 2, 4, 7, 8, 14, 28, 56
   composite 1, 2, 5, 7, 10, 14, 35, 70
   composite 1, 2, 3, 4, 6, 8, 12, 24
4. prime 1, 111
   composite 1, 3, 9, 27, 81
   composite 1, 5, 7, 35
5. prime 1, 83
   composite 1, 7, 49
   prime 1, 19
6. prime 1, 47
   composite 1, 2, 3, 4, 6, 8, 12, 16,
   24, 32, 48, 96
   composite 1, 3, 17, 51

Activity 36
1. commutative
2. distributive
3. associative
4. identity
5. commutative
6. identity
7. associative
8. commutative
9. distributive
10. associative

Activity 37
1. 3,185   7,767   3,136   3,438
2. 3,852   53,784   45,927   42,910
3. 16,070   8,064
   27,321   11,922
4. 38,120   19,686
   25,620   45,270
5. 24,738   21,303
   35,545   27,534
6. 7,434   56,152
   35,164   15,525

Activity 38
1. 142,023   23,052
   186,894   101,465
2. 35,360   110,000
   631,884   484,416
3. 882,540   351,072
   406,406   120,225
4. 62,426   252,120
   649,428   118,400

Activity 39
1. 1,608   39,865
   24,349   151,760
2. 3,222,930   3,945,000
   5,735,752   4,639,810
3. 3,136,000   20,985,220
   35,779,318   6,019,640
4. 31,070,930   3,363,000
   19,167,960   8,688,894

Activity 40
1. 0.9727   3.1042
   43.86   25.5486
2. 5.0745   32.4
   0.0027   1,714.02
3. 3.445   0.00122
   6.165   4.03
4. 0.0455   5.22
   0.159075   0.9375

Activity 41
1. 48   -48   -48   24
2. -36   -33   150   -32
3. 121   -100   -162   -17
4. -130   84   -88   75

Advantage Math Grade 7 © 2005 Creative Teaching Press

5. 400   -156   72   -840
6. -4,800   3,600   1,250   -1,700

Activity 42
1. 545   2,467 R3
   165 R3   1,952 R2
2. 535   183 R16
   613 R6   67
3. 102   56   113 R10   104
4. 30   57 R1   80 R16   42

Activity 43
1. 10,566.8   1,633
   24,690.5   3,063.75
2. 3,239.75   2,928.4
   948.75   234
3. 2,389.2   2,339.5   909.8   145.2
4. $1,379.44

Activity 44
1. 4.09   5.07   8.04   125.05
2. 25.09   56.02   37.04   13.02
3. 12.25   15.75   5.65   21.35
4. $65.85

Activity 45
1. 6.5   12.6   25.6   14.5
2. 25.6   45.3   45.8   18.5
3. 26.4   12.6   15.8   31.6
4. 7 sections
5. 136 pieces

Activity 46
1. 4   -8   8
2. -15   15.3   85.5
3. 12.6   -45.3   45.6
4. -56.2   14.6   14.6
5. -12   -24.5   -65.3

Activity 47
1. 14,558   519   66,450
   13,123
2. 3,138,623   623,162   157
   154.51
3. 692.614   86.436   -4,276.014
   4,890.134

4. 0   -7   13   1.5
5. 84   48   10   -63

Activity 48
1. 24   11.5   46   90
2. 5,880   13,068   88,450
   1,800,050
3. 12.0978   0.0048   4.992
   0.8645
4. -75   17   8   -12
5. 45   125   245.5   278.4
6. 15.6   45.6   25.3   8.5

Activity 49
1. associative
2. 4,780
3. 1,560,287
4. 28.38
5. $76.56
6. +12
7. $\frac{1}{2}$
8. 7,385

Activity 50
1. 8.69
2. 1,001.037
3. -2.1
4. 68
5. -6.75 + -0.05
6. 91
7. 99.5
8. 12

Activity 51
1. $\frac{1}{2}$ $1\frac{1}{5}$ $\frac{7}{8}$ $\frac{10}{13}$
2. $\frac{4}{9}$ $\frac{1}{2}$ $\frac{1}{10}$ $\frac{7}{32}$
3. $\frac{8}{13}$ $\frac{24}{35}$ $\frac{1}{9}$ $\frac{40}{13}$
4. $\frac{25}{64}$ $\frac{1}{12}$ $\frac{18}{19}$ $\frac{2}{31}$

Activity 52
1. $\frac{17}{30}$ $\frac{9}{20}$ $\frac{7}{12}$ $\frac{26}{35}$
2. $\frac{1}{2}$ $\frac{1}{3}$ $\frac{13}{16}$ $1\frac{5}{14}$
3. $\frac{7}{15}$ $\frac{5}{14}$ $\frac{1}{8}$ $\frac{1}{2}$
4. $\frac{11}{20}$ $\frac{5}{21}$ $\frac{11}{24}$ $\frac{23}{42}$

Activity 53
1. $-\frac{1}{5}$ $1\frac{1}{4}$ $-\frac{1}{3}$ 2
2. $-\frac{5}{12}$ $-\frac{7}{40}$
   $1\frac{17}{28}$ $-1\frac{3}{40}$
3. 2 $1\frac{3}{7}$ $1\frac{1}{4}$ $1\frac{5}{8}$
4. $2\frac{8}{9}$ $1\frac{4}{5}$ $1\frac{5}{13}$ $4\frac{1}{5}$

Activity 54
1. $1\frac{2}{5}$ 1 $\frac{7}{73}$ $-\frac{1}{3}$
2. $1\frac{1}{39}$ $\frac{13}{16}$
   $\frac{43}{326}$ $1\frac{27}{46}$
3. $\frac{17}{24}$ $\frac{19}{30}$ $\frac{9}{16}$ $\frac{25}{36}$
4. $\frac{11}{26}$ $\frac{11}{24}$ $-\frac{1}{40}$ $\frac{5}{14}$
5. $\frac{5}{18}$ $\frac{1}{8}$ $\frac{13}{14}$ $\frac{7}{15}$
6. $\frac{43}{48}$ $1\frac{7}{24}$ $-1\frac{2}{35}$ $\frac{1}{5}$

Activity 55
1. $\frac{7}{18}$ $\frac{3}{10}$ $\frac{33}{56}$ $\frac{29}{48}$
2. $\frac{3}{16}$ $1\frac{37}{72}$ $\frac{33}{80}$ $1\frac{7}{18}$
3. $\frac{1}{10}$ $\frac{2}{3}$ $\frac{7}{24}$ $\frac{13}{18}$
4. $1\frac{1}{4}$ $\frac{1}{9}$ $-\frac{2}{7}$ $-\frac{2}{3}$
5. $1\frac{1}{4}$ $1\frac{2}{7}$ $-\frac{1}{6}$ $-\frac{17}{40}$
6. $\frac{2}{3}$ $3\frac{7}{9}$ $2\frac{4}{25}$ $1\frac{3}{7}$

Activity 56
1. $\frac{1}{6}$ $\frac{1}{10}$ $\frac{1}{15}$ $\frac{1}{4}$
2. $\frac{5}{8}$ $\frac{1}{12}$ $\frac{5}{14}$ $\frac{1}{3}$
3. $\frac{4}{25}$ $\frac{2}{15}$ $\frac{3}{48}$ $\frac{5}{12}$
4. $\frac{8}{15}$ $\frac{20}{27}$ $\frac{1}{3}$ $\frac{4}{35}$
5. $\frac{5}{32}$ $\frac{2}{17}$ $\frac{55}{108}$ $\frac{1}{16}$
6. $\frac{5}{27}$ $\frac{21}{64}$ $\frac{5}{21}$ $\frac{1}{4}$

Activity 57
1. $3\frac{1}{6}$ $1\frac{1}{4}$ $2\frac{11}{32}$
2. $1\frac{1}{2}$ $2\frac{1}{2}$ $\frac{17}{48}$
3. $\frac{5}{6}$ $3\frac{3}{4}$ $3\frac{15}{16}$
4. $1\frac{1}{5}$ $1\frac{23}{40}$ $2\frac{1}{26}$
5. $29\frac{5}{6}$ $26\frac{1}{8}$ $36\frac{21}{32}$
6. $\frac{69}{128}$ $\frac{5}{12}$ $\frac{9}{20}$

Activity 58
1. $1\frac{1}{3}$ $\frac{5}{12}$ 2
2. $1\frac{1}{3}$ $1\frac{3}{5}$ $1\frac{1}{8}$
3. $3\frac{1}{3}$ $\frac{3}{4}$ $1\frac{1}{3}$
4. $1\frac{2}{3}$ $2\frac{6}{7}$ $1\frac{5}{9}$
5. 6 $1\frac{1}{6}$ $2\frac{1}{2}$
6. $1\frac{1}{2}$ $2\frac{4}{9}$ $1\frac{1}{20}$

Activity 59
1. $3\frac{2}{5}$ $3\frac{1}{3}$ $8\frac{1}{4}$
2. $2\frac{6}{7}$ $6\frac{2}{9}$ $4\frac{44}{49}$
3. $1\frac{14}{25}$ $11\frac{5}{8}$ $9\frac{3}{8}$
4. $3\frac{7}{15}$ $6\frac{3}{4}$ $5\frac{6}{7}$
5. $18\frac{1}{3}$ $3\frac{4}{15}$ 15
6. $6\frac{6}{49}$ $5\frac{1}{25}$ $6\frac{3}{40}$

Activity 60
1. -3 -5 $-3\frac{3}{4}$
2. $-2\frac{1}{10}$ $-2\frac{1}{2}$ $-6\frac{2}{5}$
3. $\frac{-5}{24}$ $\frac{-1}{24}$ $\frac{-3}{70}$
4. $\frac{-1}{18}$ $\frac{-1}{8}$ $\frac{-7}{40}$
5. $\frac{-19}{108}$ $-3\frac{1}{3}$ $\frac{-7}{60}$
6. $\frac{-1}{27}$ $-1\frac{1}{4}$ $\frac{-1}{14}$

Activity 61
1. $\frac{12}{25}$ $\frac{9}{32}$ $\frac{3}{20}$
2. 4 $2\frac{3}{16}$ $3\frac{3}{4}$
3. $\frac{5}{8}$ $\frac{1}{12}$ $\frac{9}{20}$
4. $\frac{9}{25}$ $2\frac{1}{3}$ $2\frac{2}{3}$
5. $2\frac{1}{32}$ $1\frac{4}{15}$ $3\frac{1}{20}$
6. $\frac{33}{40}$ $3\frac{3}{14}$ $2\frac{26}{27}$

Activity 62
1. $7\frac{1}{2}$ $3\frac{3}{8}$ 44
2. $3\frac{3}{8}$ $1\frac{11}{16}$ $9\frac{11}{21}$
3. $6\frac{3}{40}$ $4\frac{26}{27}$ $81\frac{1}{104}$
4. $36\frac{21}{32}$ $\frac{5}{6}$ $11\frac{1}{5}$
5. $-1\frac{4}{5}$ $\frac{-1}{8}$ $\frac{-4}{5}$
6. $-1\frac{2}{3}$ $\frac{-6}{35}$ $\frac{-1}{2}$

Activity 63
1. $\frac{2}{25}$
2. $\frac{26}{35}$
3. $2\frac{1}{2}$
4. $\frac{7}{48}$
5. $8\frac{1}{3}$
6. $1\frac{1}{2}$
7. $2\frac{19}{28}$
8. $\frac{-1}{9}$

Activity 64
1. $\frac{1}{3}$
2. $\frac{2}{15}$
3. $3\frac{3}{4}$

4. $\frac{9}{14}$
5. $17\frac{1}{15}$
6. $1\frac{2}{3}$
7. $1\frac{17}{28}$
8. $\frac{-3}{16}$

Activity 65
1. $5\frac{5}{9}$
2. 2
3. 9
4. 11
5. 30
6. -28
7. 23
8. 32

Activity 66
1. $48>33$
2. $22>13$
3. $53<79$
4. $41>11$
5. 15
6. 1
7. 41
8. -9
9. -1
10. $(7\times6) + (10\times2)=62$

Activity 67
1. 6 18 144
2. 10 36 4
3. 14 8 24
4. 3 9 $\frac{1}{3}$
5. 6 8 $\frac{2}{3}$
6. 11 7 1
7. 18 6 $1\frac{1}{3}$
8. 4 7 11
9. 6 12 18
10. 9 12 15

Activity 68
1. 15 14 1
2. -1 $\frac{1}{10}$ 2
3. 4 17 $1\frac{1}{2}$
4. -5 -13 5
5. 47 4 28

6. 8 -8 4
7. 35 26 35
8. -2 -34 14

Activity 69
1. 5 1
2. 5 13
3. 1 3
4. 3 100
5. 256 25
6. 9 4
7. 3 $3\frac{3}{4}$
8. -13 13
9. -10 12
10. 2 -1

Activity 70
1. 1 2
2. 5 6
3. 7 48
4. 7 4
5. 6 4
6. 50 72
7. 3 2
8. 6 4
9. 4 6
10. -1 -4

Activity 71
1. 12 10
2. 11 8
3. 5 2
4. 1 2
5. 7 8
6. 2 4
7. 5 1
8. -1 4
9. $-3,321$ $-2,207$
10. $-1,947$ $2,242$
11. 0.25 0.1
12. 0.04 25

Activity 72
1. $x<1$ $x<3$
2. $y>3$ $y>7$
3. $a>8$ $a<8.75$

Advantage Math Grade 7 © 2005 Creative Teaching Press

4. c<6 c>1

5. x<2 x>6

6. y>3 y>2

7. n<4 n>6

8. p<8 p>1

9. x>6 x>1

Activity 73

1. 12+ x >15 x>3

2. 3(x+2)=21 x=5

3. 1200÷4=x x=300

4. 33+8>x 41>x

5. 5x+10=70 x=12

6. 2x>50 x>25

7. Problems will vary. x=75

8. Problems will vary. x=10

Activity 74

1. 16

2. 9

3. 42

4. <

5. 6

6. 1/2

7. 14

8. -21

Activity 75

1. 8

2. 2.5

3. 4

4. 4

5. 2

6. y>7

7. 36n = 360

8. n − 27 = 195

Activity 76

1. 4:10

2. 6:10

3. 9:15 a.m.

4. 12 hours, 26 minutes

5. 1:10

Activity 77

1. -3°F

2. -18°C

3. 82°F

4. 32°C

5. 25°F

6. 14°C

7. at least 32°

Activity 78

1. 4.5 64 6.25

2. 40 6,000 0.4

3. 320 280 10

4. > < >

5. > > =

6. < = <

Activity 79

1. 141.75 20 567

2. 12 212 11.3

3. 68 453.49 110.3

4. 33.56 0.176 24

5. 906.98 440.8 -18

6. 0.353 22,046 680.58

Activity 80

1. 3 2.5 4,200

2. 2.5 10,000 51

3. 70 ½ 60

4. < > =

5. < < <

6. > < =

Activity 81

1. 9 80 5

2. 2.5 16 4,340

3. 1.5 1,500 ⅜

4. < > >

5. < = >

6. > < =

Activity 82

1. 1,608.2 8 1.06

2. 1.524 21.14 1.6

3. 5,150 83.82 2

4. 0.9 20 3.5

5. 37.84 1.59 2.114

6. 3,784 4.7 2.6

Activity 83

1. ∠ E, ∠ DEF; obtuse

   ∠ B, ∠ABC; right

2. ∠ N, ∠ MNO; acute

   ∠ Y, ∠XYZ; reflex

3. Angles will vary (should be three).

Activity 84

none — ∠HAF 90°

∠ DAE 25° ∠ IAF 115°

∠ CAD 30° ∠ CAF 120°

∠ GAF 16° ∠ GAD 106°

none would be 9° ∠ IAG 99°

Activity 85

1. 12 cm 16 m 3 in.

2. 16 in. 14 in. 12 in.

3. 24 ft. 18 cm 6 cm

4. 40 in. 16 yd. 6 ft.

Activity 86

1. 3.14 in. 31.4 in. 15.7 cm

2. 6.28 ft. 18.84 cm 37.68 in.

3. 43.96 mm 11.775 in. 13.188 yd.

Activity 87

1. 50° 60° 110°

2. 40° 35° 80°

3. 26 in. 16 cm 17 ft.

Activity 88

1. 49 sq. cm 64 sq. in. 16 sq. yd.

2. 12 sq. in. 24 sq. ft. 16 sq. ft.

3. 196 sq. ft. 36 sq. cm 40 sq. in.

Activity 89

1. 153.86 sq. cm 50.24 sq. in.
   314 sq. in.

2. 7,850 sq. mm 176.625 sq. cm
   113.04 sq. in.

3. 200.96 sq. cm 28.26 sq. in.
   3.14 sq. ft.

Activity 90

1. 25 sq. in. 18 sq. cm 12 sq. cm

2. 35 sq. cm 22 sq. ft. 30 sq. in.

Activity 91
1. 16 sq. cm   60 sq. in.   105.12 sq. in.
2. 147 sq. cm   90 sq. yd.

Activity 92
1. 108 sq. ft.   72 cu. ft.
2. 214 sq. cm   210 cu. cm
3. 79.28 sq. ft.   44.928 cu. ft.

Activity 93
1. SA=36 sq. in.   V=12 cu. in.
2. SA=141.86 sq. ft.   V=90 cu. ft.

Activity 94
1. SA=1,099 sq. mm   V=2,355 cu. mm
2. SA=140.1225 sq. in.
   V=56.71625 cu. in.

Activity 95
1. 1:40
2. about 28 g
3. about 3,200 m
4. about 1 L
5. obtuse
6. 190°
7. 25°
8. the other right angle

Activity 96
1. 17 ft.
2. 3 times
3. 15 in.
4. 19.72 in.
5. It is obtuse.
6. 20 cm
7. 50.24 sq. in.
8. 34 sq. cm

Activity 97
1. 54,495,031
2. 8.76×10^5
3. -32
4. 7
5. 0.000083
6. 10^2
7. 0.625
8. 600

Activity 98
1. 3,305
2. 2.375
3. 5,009
4. -8
5. associative
6. 65.5
7. -21
8. 18

Activity 99
1. $5/12$
2. $7/12$
3. 3
4. 11
5. $-2/9$
6. 4
7. $3,312 - x < 80$ ft., $x < 3,232$ ft.
8. 8:20

Activity 100
1. 15 cm
2. 12.5 cm^2
3. 3 ft. x 3 ft. = 9 sq. ft.
4. 40 degrees
5. 10 yd.
6. $9 \text{ m}^2 \times 3.14 = 28.86 \text{ m}^2$

Advantage Math Grade 7 © 2005 Creative Teaching Press